No. 617
$8.95

HOW TO REPAIR SMALL GASOLINE ENGINES

by Paul Dempsey

TAB BOOKS
Blue Ridge Summit, Pa. 17214

FIRST EDITION

FIRST PRINTING—OCTOBER 1972
SECOND PRINTING—SEPTEMBER 1973
THIRD PRINTING—DECEMBER 1973
FOURTH PRINTING—FEBRUARY 1974

Copyright ©1972 by TAB BOOKS

Printed in the United States
of America

Reproduction or publication of the content in any manner, without express permission of the publisher, is prohibited. No liability is assumed with respect to the use of the information herein.

Hardbound Edition: International Standard Book No. 0-8306-2617-4

Paperbound Edition: International Standard Book No. 0-8306-1617-9

Library of Congress Card Number: 72-87454

Preface

This book covers many types of small gas engines, those ubiquitous 2-cycle and 4-cycle, single and multicylinder engines seen in fixed industrial use, on lawnmowers and power saws, on go-karts and snowmobiles, and on boats, both inboard and outboard. I have included just enough theoretical material (words, not math) so that the related troubleshooting procedures would be useful not only to the professional mechanic but of great utility to the interested general public, especially the do-it-yourself home handyman.

Except for Chapter 1, which gives an overall view, and Chapter 8, on business practices, each Chapter is arranged so as to reflect the "frequency of repair" analyses that have been compiled over many years. For example, ignition problems (Chapter 2) come before fuel problems (Chapter 3) because surveys have revealed that ignition is cited in 87 percent of customers' trouble complaints. Within each Chapter (from 2 through 7), in addition to the obvious troubleshooting and repair instructions, I have included practical tips and hints to help you get the job done correctly and speedily. All suggestions are based on years of practical experience.

Finally, in the last Chapter, I have included many ideas about starting up and running a repair shop. That Chapter is last deliberately—anyone thinking about going into repair work professionally should first read the entire book, and then take a cold, calculating look at Chapter 8.

I want to thank the firms which have been so kind in providing information and illustrations for this book. It would not have been possible without their cooperation.

I also want to thank Mr. Marcus Meadows for his advice on outboard motors and Professor J. San Martin for his aid in compiling the chapter on electrical systems. And finally, this book owes a great deal to Mr. Donald Keyes, small engine shop foreman at Foley's Department Store, Houston, Texas. Mr. Keyes taught me the trade. "Do it right" is his motto.

Paul Dempsey

Contents

1 **Theory of Operation, and Troubleshooting Principles**   7
Gas Engine Basics—Operating Principles—Overall Troubleshooting Principles—Troubleshooting Specific Symptoms—Summary

2 **Ignition Systems**   47
Spark Plugs—High Tension Cables—Breaker Points—Condensers—Distributors—Battery Ignition System—Magneto Ignition Systems—Special Ignition Systems

3 **Fuel Systems**   85
General Troubleshooting Procedure—System Troubles—Carburetors—Fuel Tanks, Leaks and Repair

4 **Electrical Systems**   123
Electrical Theory—Direct Current Generators—Starter Motors—Starter Generator Combinations—Alternating Current Systems—Voltage and Current Regulation—Fuses and Circuitbreakers—Lead-Acid Batteries—Wiring

5 **Engine Service**   157
Cylinder Heads—Valves—Push Rods and Rocker Arms—Pistons—Rings—Cylinders—Connecting Rods—Crankcase—Crankshaft—Inserting a Heli-Coil

6 **Power Transmission**   205
Shaft Drive—Friction Drive—V-Belt Drive—Adjustment and Replacement—Chain Drive—Gear Drive—OMC Electric Shift—Transaxles—Kickstarters—Clutches

7 Accessories and Controls   **235**

Rewind Starters—Impulse Starters—Remote Controls—Governor—Compression Releases—Pumps—Thermostats—Radiators—Oil Coolers—Mufflers

8 So You Want to Start a Shop   **269**

Work Area Layout—Safety in the Shop—Customer Relations, and Transaction Records—Complaints—Shop Tools and Supplies—Nonspecialized Shop Tools—A Final Word

Index   **283**

Chapter 1

Theory of Operation, and Troubleshooting Principles

The actual repair of small engines is not difficult. The problem lies in pin-pointing what is wrong. Later in this chapter, I will outline a simple, step-by-step procedure which can help you isolate the problem. Behind this procedure is what one might call the basic law of gas engines: an engine will run if it has ignition (at the right time), air and fuel (in the right amounts), and compression. But before we go into general and specific troubleshooting methods, a short review of the basic principles of gas engines is in order.

GAS ENGINE BASICS

In the following discussion, only simple arithmetic is used to illustrate basic truths.

Bore and Stroke

The most basic classification of any gasoline engine is in terms of its bore and stroke. The bore is the diameter of the cylinder, and the stroke is the distance the piston moves from bottom dead center (BDC) to top dead center (TDC). Bore and stroke can be expressed in inches, as is the American practice, or in millimeters. When the bore is multiplied by itself (Bore^2) times the stroke times 0.7854 times the number of cylinders, we have the displacement or the total swept area of the engine. For example, the single-cylinder Tecumseh Model D has a bore of 2½" and a stroke of 1 11-16". Applying the formula Bore^2 x Stroke x 0.7854 x number of cylinders gives a displacement of 8 cubic inches. (The actual number is 7.9766+.) As a very rough rule of thumb, subject to hundreds of exceptions, the greater the displacement, the more powerful the engine will be.

Compression Ratio

This numerical ratio is a comparison of the total swept volume of any given cylinder to the total volume of the cylinder. There will always be some area in the combustion chamber above the crown of the piston when it is at top dead center. The higher the compression ratio, that is, the more the fuel-air charge is squeezed together, the greater will be the energy released. But there are practical limits to increasing the compression ratio (CR) beyond 9 or 10 to 1.

The major problem is that high compression increases the temperature of the internal parts of the engine beyond the capacity of the cooling system. The tip of the spark plug, or the edges of the valves, can become so hot that they act as glow plugs, and ignite the mixture before the piston reaches TDC. This pre-ignition tends to drive the piston backwards, against the direction of rotation. More and more heat is generated until the engine destroys itself.

Another problem associated with high compression ratios is detonation or "spark knock." Under heavy loads and slow rotational speeds, the normal combustion process, which is really a very rapidly moving flame front, is replaced by particularly violent explosions. If continued, detonation can break pistons and cylinder heads.

We have looked at the classic method of computing compression ratio. But a little thought will show that this is not accurate when the valve and port timing is taken into consideration. The formula assumes that the cylinder bore is closed to the atmosphere. Actually, the valves or ports (on two-cycles) are open much of the time that the piston is on the up stroke. The effective compression ratio is much lower than the formula would indicate. Most Japanese manufacturers now give the effective ratio in their tables of specifications, while the Europeans are staying with the old formula. Thus, a comparison of the stated compression ratios of a Suzuki and say, a Spanish Montesa, will be misleading.

Horsepower and Torque

These two concepts are interrelated but distinct. One horsepower is the force required to lift 550 pounds 1 foot in 1 second. Note that horsepower refers to work done over time. A 10 HP chain saw should be able to do more work than one which only develops 5 horsepower. Torque, on the other hand, refers to the twisting motion on the crankshaft. It is measured in foot-pounds; one foot-pound is a force of one pound exerted on a one foot lever. Torque, therefore, is a measurement of instantaneous force and does not contain any concept of time. You can understand the distinction between these two terms with the following example: two vehicles weigh the same and have identical horsepower and gearing. Both ought to reach the same top speed. But the one which develops greater torque will reach that speed sooner. In industrial applications, torque shows up in the ability of the engine to accept loads without bogging down and losing rpm.

Torque can be related to horsepower by assuming that the twisting force acts on the end of a one foot crank. The force would act over a distance of 6.28 feet—the circumference of a circle with a radius of one foot—with each revolution of the shaft. If you know the load, and the rpm, you can easily calculate the foot-pounds of work per second. To get the horsepower, multiply the rpm by the torque and divide by the constant 5252.

Slide rule calculations enable engineers to estimate horsepower with some accuracy, but the test is to mount the engine on a dynamometer, a device which can accurately load the engine. The load may be a friction brake, a generator, or a water turbine. The latter two are most commonly used because of their precision. Normally, the engine is mounted in a flexible cradle connected to a spring scale. As the crankshaft turns, it will exert an equal and opposite force, twisting the cradle in the direction that is counter to the direction of crankshaft rotation. Torque is then read directly from the scale.

The load is adjusted to hold the engine at a given rpm with the throttle open. Then the load is reduced and the engine speed is increased to the next test point. Fifteen or twenty tests across the rpm scale provide enough data to plot torque and horsepower on graph paper as a curve.

Fig. 1-1 shows typical torque and horsepower curves of three Kohler industrial engines. Note that while the torque curve of these particular engines begins to fall off at 2,400 rpm, the horsepower curve continues upward. It may be asked: how is this possible since torque times rpm gives horsepower? The answer is that so long as torque is falling off at a rate less than the rate of rpm increase, horsepower will continue to climb. The horsepower curve will peak at the point where torque falls off in a one to one relationship to rpm increase.

The curves shown in Fig. 1-1 are relatively flat, which is typical of industrial engines. These engines must pull from low rpm and are expected to operate over a fairly wide rpm range. Highly-tuned sports engines characteristically have peaky torque and horsepower curves which attain their maximum values at high rpm. For example the Suzuki T20 "Super Sport" motorcycle power plant attains its maximum horsepower at 8,000 rpm and its best torque at 6,500 rpm. Pure racing engines

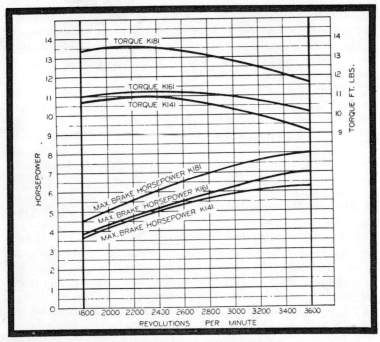

Fig. 1-1. Horsepower vs Torque Graph. Note that the torque curves are flat, showing that these engines have good "pulling" characteristics across rpm range. (Courtesy Kohler of Kohler.)

may operate most efficiently at double this rotational speed.

As a general rule, torque is greatest at medium engine speeds. Friction and reciprocating losses are low and the engine has time to "breathe" a full charge of gasoline and air. As the speed increases, the friction load generated by the rings, piston, and bearings increases dramatically. In most cases friction goes up as the square of the rpm: for example, at 9,000 rpm, the friction will be (3^2) or nine times what it was at 3,000 rpm. The same is true for reciprocating loads—the energy to move the piston and the valve up and down also increases as the square of the speed. The carburetor, induction tract, and muffler also lose efficiency as the gas flow velocity increases.

Larger engines, bigger parts, and more and heavier reciprocating masses take their toll. This is why model airplane engines often develop 3 HP or even 4 HP per cubic inch while a 9 cubic inch Briggs and Stratton is rated at only 3½ HP. The problem can be solved at least partially by going to more cylinders. The last 125 cc Honda racing motorcycle had five cylinders, each no larger than a thimble.

But while small engines (or cylinders) can develop more horsepower per cubic inch, they are limited in torque. The larger engine will produce more torque simply because it burns more fuel and oxygen per cycle. There is no substitute for cubic inches—at least where mid-range performance is concerned.

Number and Placement of Cylinders

The first decision an engine designer must make is: based on the expected usage, how many cylinders must this engine have? While most small gas engines are singles, some have contained as many as eight cylinders. The single-cylinder engine is mechanically simple, inexpensive to manufacture, and easy to service in the field. Good torque is also a characteristic of most singles. The piston area is larger than on an equivalent multi-cylinder engine, thus giving more surface for the explosion to push against. And, in most cases, the stroke will be longer, giving a better mechanical advantage.

But there are disadvantages to the single as well. There is no fail-safe capability. On a multi-cylinder engine, however, if a spark plug, or a piston, or a valve fails, the engine will still run. The mechanical complexity of a multi increases its reliability. And this complexity is not as expensive as it might look. Once the machine tools have been set up, it costs little more to make four to six cylinders rather than one. The multi also comes out ahead on the basis of power pulses per revolution. In extreme performance situations, the multi uses its smaller piston area to an advantage. Less area at the crown of the piston means that there will be relatively more area along the sides on the piston skirt. The skirt transfers heat from the crown to the cylinder walls. The more area in contact with the walls, the more heat will be transferred, and the less likelihood of "holing" a piston or a cylinder wall.

Small industrial engines are classified according to the plane of the crankshaft relative to the ground. **Vertical** shaft engines are used on most rotary lawnmowers. The end of the crank provides a convenient mounting point for the blade. **Horizontal** engines have the crank parallel to the ground and are used on reel type mowers, power tools, and go-karts. Other than the position of the shaft and the carburetor float, these two styles of engine are similar.

Classification is also by the placement of the cylinders. A horizontal-crank single with the cylinder inclined is often called a "sloper." Continental and Reo have used this pattern

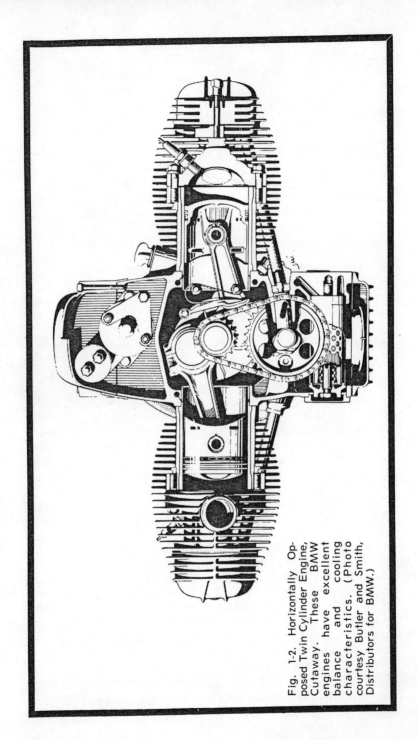

Fig. 1-2. Horizontally Opposed Twin Cylinder Engine, Cutaway. These BMW engines have excellent balance and cooling characteristics. (Photo courtesy Butler and Smith, Distributors for BMW.)

for their small industrial plants. The main advantage seems to be compactness. Horizontally-opposed engines are multis with the cylinders 180 degrees apart. A famous example from the automotive world of "pan cake" configuration is the Volkswagen. Some Power Products two-cycles, the Kohler K482 and K662, and the current range of BMW motorcycles, are horizontally opposed (Fig. 1-2). These engines are compact, easy to cool, and have good inherent balance. The vee type is similar, except that the angle between the cylinders is smaller. Most vee's are built with an included angle of 90 degrees between the cylinders, although a few have been as narrow as 60 degrees (Fig. 1-3). A little thought will show that the firing impulses on a twin cylinder vee are uneven (which

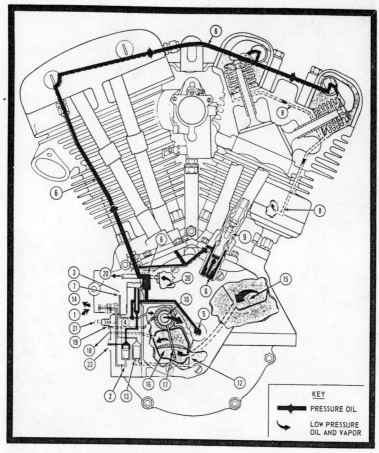

Fig. 1-3. Vee-Type, Twin-Cylinder, 74 Cu. In. Harley-Davidson Engine, Cutaway showing lubrication system. (Courtesy Harley-Davidson.)

Fig. 1-4. Vintage 4-Cylinder In-line Nimbus Motorcycle Engine with Single Overhead Camshaft. (Photo by Al Galinski.)

accounts for the loping idle of the big Harleys), but this is not noticeable at operating rpm. The vee configuration is also found on some of the large outboards. Both the horizontally opposed and the vee types use a single crank throw for two cylinders. Looking down on the engine, you'd see that the bores are off-set by the width of the rod. The short, stubby crankshaft and the rectangular block resist distortion.

Another configuration which should be mentioned, if only in passing, is the "square four." Pioneered by European aircraft designers, square fours came into the United States on Ariel motorcycles. These engines had four cylinders and two crankshafts, laid side by side. The cranks were connected by a pair of gears. Good balance was achieved, since each half of the engine turned the opposite direction; but the design was complicated and expensive to maintain.

Currently, the most favored configuration for small, multi-cylinder power plants is the in-line engine, examples of which are shown in Figs. 1-4 and 1-5. Typical examples are the Honda multis, the BSA and Triumph twins, and Johnson and Mercury outboards. Advantages are ease of manufacture, buyer acceptance (because of automobile practice, people seem to accept this configuration as "natural"), and the excellent balance afforded by threes and sixes.

In-line engines are further classified as to the position of the block in the frame. The engine may lie transversely or longitudinally. On liquid-cooled designs the lay of the engine is a matter of design convenience, but on motorcycles (air-cooled engines), the transverse position has an important advantage. All the cylinders are in the air stream and are cooled equally. The Nimbus pictured in Fig. 1-4 suffers from overheating of the rearmost cylinders.

Classification of Engine by Speed

On the basis of the torque and horsepower curves, engines can be divided into high- and low-speed types.

Low-speed engines are designed to be operated in the range of 3,000 to 4,000 revolutions a minute. Usually of heavy, cast iron construction, these workhorses drive pumps, generators, and the like. High-speed engines are "red-lined" at anywhere from 7,000 to 11,000 rpm and are used where engine weight is a critical factor. Some of these engines develop better than one horsepower per pound. High-speed engines are used in power chain saws, go-karts, and some of the faster motorcycles. It is generally believed that high-speed engines are less durable than their low-speed counterparts.

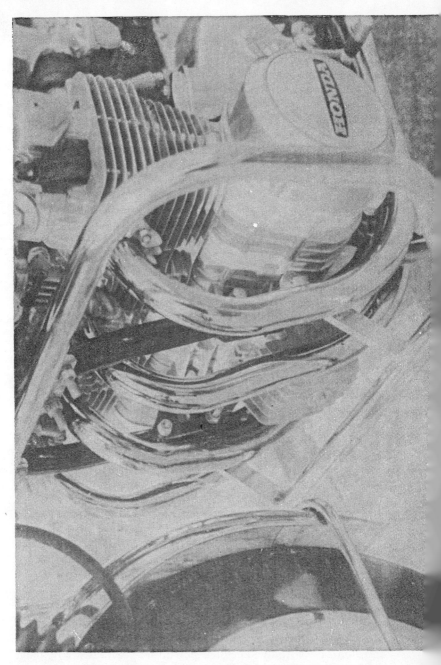

Fig. 1-5. Four-cylinder, In-line DOHC, Transverse-Method Honda Motorcycle Engine. Extensive use of aluminum alloy, and transverse mounting assures cool operation. (Photo by Al Galinski.)

Years ago, in the days of babbitt bearings and cast iron pistons, this may have been true. But today, thanks to modern design techniques (short strokes and smaller, lighter reciprocative masses), close manufacturing tolerances, and advances in metallurgy, high-speed engines are as long-lived as any.

Vibration

Whenever a force displaces an elastic body from its position, the body will develop a restoring force which acts to return it to its original state. Owing to the inertia of the body, the return movement will carry it beyond the original position. A restoring force is developed which repeats the cycle in the opposite direction. The vibration will continue until it is dampened by friction.

Every elastic body, and this term includes objects such as crankshafts and pushrods, has a natural frequency of vibration, much like a tuning fork. As a matter of fact, some of the old timers tested cranks for flaws by striking them with a hammer. A good crank will have a particular ring; it will be muted if the crank is cracked. The natural frequency phenomenon can also lead to trouble. If the forced vibrations occur at the same frequency (or a multiple of the frequency) as the natural vibration, the amplitude will increase. If you have a tuning fork which vibrates at 360 cycles a second and arrange to strike it 360 times a second in the synchronous plane, the tines will displace further with each blow until the fork destroys itself. Sometimes this happens on overhead valve engines. The valve gear—tappets, push rods, rocker arms, springs, and the valve itself—has a natural frequency. If the rpm is such that the forces acting on the mechanism are the same as the natural frequency, tremendous stresses occur and failure is imminent.

Vibration in piston engines is caused by imbalance, and because of the nature of the beast, cannot be eliminated entirely. Multi-cylinder engines generally have better balance than do singles, since the pistons can be arranged to "cancel" each other out. The crankshaft should be as short and as wide as possible and supported by bearings between each cylinder, on inline designs, or between each pair of cylinders, on opposed or vee-types. In special situations, a counter-rotating shaft can be employed, which will limit the tendency of the engine to react against its mounts. Briggs and Stratton, and Ford of Germany, have used this method. Another approach is to isolate the engine by means of rubber mounts.

Noise

Wherever there is vibration, there is noise. A few years ago, hardly anyone was concerned with the noise made by small engines, and if you couldn't nap because your neighbor was mowing his lawn, you cussed and put a pillow over your ears. But things are different today. The proliferation of small engines into all corners of the environment—the lakes, the desert wilderness, the winter forests, and into the heart of the cities—has aroused national attention on the noise problem. Chicago already has a noise ordinance on the books which, if enforced, would effectively ban small engines from the city. The Federal Government is expected to follow suit.

Why is it that one power chain saw makes more noise than 50 automobiles? The answer is complicated. It is not that small engine manufacturers want to damage the hearing of their customers (studies have shown that hearing loss, especially for high frequencies, has occurred among chain saw operators and motorcycle racers), or invoke the wrath of the government. Small engines have a high specific output when compared to automobile engines. The little engines work harder and so are inherently noisier.

The problem is compounded by the fact that single- and twin-cylinder mufflers are not as efficient as those designed for multi-cylinder units. In any exhaust system, there are alternating high pressure and low pressure waves. When the exhaust valve or port opens, it releases a high pressure wave which carries a near vacuum in its trail. Automobile engineers have learned to combine these low and high pressure waves to give a low average pressure. Consequently, the level of sound is reduced. In theory this could be done with singles and twins, but the plumbing would be as large as the engine. Instead, small engine designers have to be content with baffles (which slow down the exhaust gases and reduce their sound-making potential) and packing, which absorbs some of the high frequency sounds.

But even with a dead silent muffler, small engines are still noisy. Recently Tecumseh ran a series of tests on a 3½ HP four-cycle lawnmower. The results are interesting. At 3,225 rpm the engine produced a sound level of 68 db A. The term "db A" refers to decibels measured at fifty feet (the "spectator distance" for most power equipment) and sound consisting of frequencies which are intrusive on human hearing. With the exhaust completely muffled, the engine still produced 65.3 db A. Since the decibel scale is logarithmic, small changes are significant—most people can perceive a sound variation of

1 decibel—but this test showed that the muffler is only a small part of the problem. Other tests have shown that larger engines are even less responsive to increases in muffler efficiency.

Much of the noise is generated in the induction tract. Air whistles through the filter and carburetor and, on two-cycle engines, the reed valve makes a distinct sound as it slams shut with each revolution of the crank. Modern engines now have more "hiss felt" lining the intake tract. This material absorbs sound above 2,000 Hz. (Hz = Hertz = cycles per second.) Another approach is to use fiber reeds in valves rather than the spring-steel type.

Quiet mufflers and quiet intake tracts help, and they can be achieved in the present state of the art. Another, knottier problem, is the noise made by mechanical parts. The cylinder head vibrates with each explosion. On an air-cooled engine, the amount of sound transmitted in this way goes up as the cube of the bore diameter. For example, triple the size of the cylinders and the noise factor goes up nine times. Another culprit is the operating clearances on reciprocating parts. Noises caused by piston slap (the side to side movement of the piston in the bore), valve lash, and the clearances between the crankpin and the connecting rod bearing, all reinforce each other. Timing gears are also noise makers. The flywheel and crankshaft pulley pick up vibrations from the crankshaft and radiate them outward in the manner of a "tin can telephone" such as children play with. Any large sheet metal surface such as the oil sump or the cooling shrouds also has this effect. On riding lawnmowers, it is not unusual for the whole frame to serve as a sounding box.

But progress is being made. The 1972 Mercury "650" is one of the quietest outboards ever built. It is the result of years of research in an anechoic chamber (this is a sound proof room with the inner surfaces arranged so that there will be no echoes).

Mercury engineers attacked all three areas: the exhaust, the intake, and mechanical noise. A new intake system was designed and the exhaust was modified to reduce the organ pipe effect. Mechanical clearances were tightened and the power head was encapsulated in a sound-proof cover. As a result, the db A level was cut in half.

Cooling

This section on cooling, and the section on lubrication which follows, are related because oil is also a coolant.

The best gasoline engines are only about 40 percent efficient. Only a little more than half of the heat released by the fuel is used to make motion. The rest of the heat is wasted in friction, transferred to the pistons and cylinder walls, and expelled through the tail pipe. The temperatures involved here are high: combustion takes place at about 2,000 degrees F. The exhaust gases range between 1,000 and 1,200 degrees and the crown of the piston often reaches 1,000 degrees F. Without some method of taking heat out of the engine and bearings, these parts and then the engine itself would melt. At only 350 degrees F, white metal bearings (consisting of tin and lead) begin to soften.

Liquid-cooled engines use water as the medium of heat transfer. The water circulates in a double-walled chamber, called a jacket, which surrounds the combustion chamber and the cylinder bores. The water jacket transfers heat from these hot spots to the atmosphere. Most liquid-cooled engines use a radiator which is equipped with an engine-driven fan and may have an expansion tank to prevent fluid loss. Outboards and some inboard marine engines dispense with the radiator since they draw upon an unlimited supply of cool water.

Much research has been done on the effects of different coolant temperatures. At one time it was believed that moderate outlet temperatures were desirable since the formation of scale is retarded at below 180 degrees F. Scale is a rock-like combination of minerals from the water and rust or aluminum oxide. It builds up in the water passages until, in extreme cases, the water supply is cut off entirely. And even minute amounts of scale act as insulation.

However, contemporary theory favors much higher outlet temperatures, say 220 to 250 degrees F. At these temperatures, there is less of a tendency for the crankcase oil and the exhaust gases to combine and form acids. Consequently, bearing and bore life are increased. And higher temperatures mean that the oil flows easier, thus reducing the amount of energy wasted in friction. Another plus is that because of the greater differential between the water and air temperatures, a smaller radiator and fan assembly can be used. But scale remains a problem and these engines must be periodically inspected.

Marine engines with open systems, such as outboards, can require considerable water jacket maintenance. In fresh water, mud and marine growth can block the passageways. In salt water, the problem is compounded since salt attacks cast iron and aluminum. The engine does not have to be running for the damage to occur. Dealers tell of engines rusting out on the showroom floor. Whenever an engine of this type has been operated in salt water, it must be flushed with clear, fresh water. And in no case should the outlet temperature be more than 160 degrees F. Otherwise salt will drop out of solution and cake in the jacket.

On closed (fresh-water) systems, it is advantageous to use 50-50 or even a 75-25 mix of ethylene glycol and water. This increases the boiling temperature of the coolant and, of course, lowers the freezing point. Ethylene glycol anti-freezes sold by the major suppliers, such as Union Carbide and DuPont, also contain rust inhibitors and water pump lubricants.

Most liquid-cooled engines use a pump to circulate the water, although a few still employ the thermosiphon method which Henry Ford chose for his Model T. Cold water is denser than hot water; circulation will result if the water is heated at one point and cooled at another. Thermosiphon systems use a hot water line that runs from the cylinder head to the top of the radiator. Needless to say, circulation is rather leisurely, and this method is not used on high performance engines.

Air-cooled engines dissipate heat by means of fins cast on the exterior cylinder wall and cylinder head. The more fin area, the more heat is transferred. Current practice is to use large fins with a thin cross-section and sharp edges. Heat will "bleed" more rapidly from a sharp surface than from a blunt or rounded one. But fins alone are not enough. There must be a moving current of air over the fins. Motorcycle designers assume that the vehicle will be in motion. It is possible to overheat a motorcycle by long periods of idling. Therefore, when doing extended work on a stationary motorcycle, it is advisable to use a large, high velocity fan to force-cool the engine.

Industrial engines have a built-in fan, which is usually cast as part of the flywheel. Air is pulled in around the center of the hub and is expelled around its edges where it is taken (ducted) over the engine through a shroud which encloses the engine.

Air-cooled engines reach their operating temperature quickly, a factor which reduces piston and bearing wear. And because air cooling is the simplest system, it is quite reliable. (A water pump which you don't have cannot fail.) About the only exterior maintenance which is required for an air-cooled engine is an occasional cleaning of oil and dust accumulations from the fins and the shroud.

On the other hand, there are disadvantages to air cooling which have prevented this system from becoming universal.

These engines are hot to the touch, and this can be a handicap in some applications. Noise is another problem—the fins which radiate heat also radiate sound. Shrouding might muffle some of the sound, except that the shrouds themselves vibrate and act as sounding boards. And in extreme performance situations, air cooling simply cannot handle the heat. No serious Indianapolis contender is air-cooled.

Lubrication

Two-cycles are lubricated by oil mixed with fuel. Relatively small amounts of oil are required—some engines can run as lean as 100 parts of gasoline to one part of oil. Until recently, all two-cycles required that the owner mix the oil with the fuel in the correct proportions. The drill was messy and wasteful. Some owners felt that if a little oil was good, more would be better. And lubrication requirements vary with rpm and load; the 100 to 1 figure quoted above is specified for full throttle operation. At idle, these engines have more oil than they need. The excess accumulates in the exhaust ports as carbon or is expelled into the atmosphere. Japanese engineers made a considerable breakthough with the auto-lube system. Essentially the auto-lube is a variable displacement pump controlled by the position of the throttle. Oil is delivered as the engine needs it in proportions which vary from 400 to 1 at idle to 30 to 1 on acceleration. If the government moves to control air pollution created by small engines (there have already been some hints from Washington that this is in the offing), some form of auto-lube will be mandatory on all two-cycles.

Four-cycle lubrication systems are divided into two broad classifications. The simplest is the splash system. Oil in the crankcase is splashed around the moving parts as the crankshaft throws come down. Sometimes there will be a "scooper" on the end of the connecting rod or (on Briggs and Stratton designs) a vaned wheel which revolves with the crank. The splash system is adequate for low-speed engines, but has serious drawbacks when the rpm goes up. In the first place, the engine must be an L-head; overhead valve gear is far too remote to receive oil. At high rpm, bearing lubrication is a problem. And there is the power loss caused by "windage," the friction of oil against the throws.

A better system is to employ an oil pump. Most oil pumps are the geared type. On most engines, the pump consists of two meshed gears turning in a closed housing. Oil is picked up on one side and is forced out the other. Tecumseh engines are somewhat unusual since they employ a reciprocating pump and check valve arrangement. In either case there is a spring-loaded relief valve to keep the pressure from exceeding preset limits.

Refer again to Fig. 1-3 which shows the oil circuits of a typical Harley-Davidson engine. Note that every friction surface receives oil, either under pressure or by gravity on the return to the crankcase. The Harley uses a dry sump, that is, the crankcase is dry and oil is stored in an external tank (15). The oil pump is really two pumps in one; one side pressurizes the engine and the other side evacuates the crankcase. The pump which discharges the crankcase has a greater capacity than the pressure pump; this keeps the case from flooding. Dry-sump lubrication has certain advantages for high performance applications. Since the crankcase is dry, there can be no power loss due to "windage." The external tank, mounted well away from the engine block, tends to act also as an oil cooler. And a dry-sump engine can be made with a low center of gravity, which aids handling. The price is additional complexity and external oil lines which are subject to leakage.

Motor oil has four main functions. The first is to reduce friction by interposing a film between rubbing parts. This film may only be a few thousandths of an inch thick, but as long as it remains intact it will prevent wear and heat build-up. Most of the wear on the typical engine occurs on cold starting. The oil has drained off the parts, and for the first few revolutions, there is no lubrication to speak of. When rebuilding an engine, mechanics use liberal quantities of oil on all friction surfaces. On four-cycles, some even go so far as to pressurize the oil galleries before the engine is started for the first time.

Another function of oil is to act as a seal between the piston and the bore, and between the valves and their guides. The sealing effect is considerable, especially when the engine is worn. A major reason why engines which have been inoperative for a long period are difficult to start, is that the oil has drained from the cylinder bore.

A third function is to remove heat. In this sense, all engines are liquid-cooled. The figure is considerably less for a two-cycle engine, but on a typical four-cycle, some 60 percent of the cooling chore is handled by the oil. This is why the crankcase oil level is so critical. Engines which run low on oil fail, not because of lack of lubrication—there is usually enough for that—but because of heat build-up. Small quantities of oil cannot dissipate the heat quickly enough. Some four-cycles employ an oil radiator to help keep temperatures below the critical 250 degrees F for 30 W oil and 300 degrees F for 50 W.

Finally, oil must resist and neutralize the harmful byproducts of combustion. A gallon of gasoline passing through the combustion chamber will combine with oxygen to release approximately one gallon of water, up to (depending on the engine) 0.1 gallon of unburned gasoline, and traces of hydrochloric, hydrobromic, and various nitrogen- and sulfur-based acids. There will also be small quantities of varnishes, resins, soot, and lead salts. Most of these chemicals are expended out the tail pipe, but some of them get past the rings and are trapped in the crankcase. One study has shown that most bore wear is not caused by friction, as one would suppose, but is caused by acid etching. Sludge accumulates around the piston rings and causes sticking. In extreme cases, sludge can block off oil passages and fill the screens on oil pumps. Water causes rust, and the unburned fuel dilutes the oil, increasing the wear factor.

In order to take some of the load off the oil and to extend oil change periods, more and more small engines are employing some form of oil filter. The filter element, which is usually pleated paper (although other organic materials such as cotton stuffing have been used), is designed to trap solids. Modern filters can trap particles as small as 25 microns in diameter (a micron is one millionth of an inch). A good filter will reduce wear and will localize damage if one part begins to pulverize. Fortunately, most small engine filters are in series with the pump—all oil going to the engine must pass through the filter. As a safety factor, if the filter is stopped up, a pressure relief valve will open, assuring full lubrication as the oil bypasses the filter. A few of the older engines used mostly on construction work and in marine applications had bypass filters. These filters were in parallel to the main oil line. Sooner or later all the oil would be filtered, but in the meantime loose particles were free to circulate around the engine.

Regardless of advertising claims, no filter can remove liquid contaminates. To do so, the filter would have to be a miniature oil refinery.

All motor oils are rated by viscosity, which is defined as the resistance of a fluid to the force that causes it to flow. The higher the number, the thicker, or more viscous, the oil. Thus SAE (Society of Automotive Engineers) 40 has more resistance to flow than does SAE 5 W. The "W" suffix means that the oil has been tested for winter operation. The industry has had to provide different weights because oil tends to lose its viscosity as the temperature goes up. 5 W is excellent for sub-zero operation, but would thin to uselessness in the Mohave Desert. In the last decade, multi-grade oils have been developed—such as 10 W-40—which hold their viscosity over a wide range of temperatures.

In addition to viscosity, motor oils are classed by the American Petroleum Institute for load strength, acid neutralization, resistance to sludge build-up, and rust inhibition. Because of ever-extended oil change intervals for new American cars, the rating system is in a state of flux. The rating system is based on automotive use but is applicable to small engines. Hence, do not use less than API Service MS ("Severe Service" under the old system) in any engine. Under the new rating, devised for 1971 cars, use no lower quality than API SD. The best oil available at this writing is API SE. It is far superior to either the MS or SD grades, and its use is perhaps justified in new engines, operated under the most extreme conditions.

Oils are also classified as detergent and non-detergent. Detergent oils contain dispersants which keep sludge and carbon particles in suspension. All manufacturers recommend the use of detergent oils—in fact, with multigrades there is no choice in the matter. The only caution is that when used in an old, dirty engine, a detergent oil can loosen the accumulated sludge to the point where oil passages may be blocked.

Switching to different brands of oil is frowned on by many mechanics, but there is no evidence that the practice is harmful. The military has been doing it for years. However, it is wise to stay with the major refiners; there is no policing of API classification numbers, and some of the "cheapie" oils are not what they are claimed to be. Oil additives should be avoided. Some, especially those that contain metals, are positively harmful. Others, while they can improve poor grades of oil, are more expensive than good oil to start with.

While all small engine makers suggest a major brand name and that API MS quality or better be used in their products, there is some difference of opinion on the grade. For the Kohler twin cylinder industrial four-cycles, the specifications are as follows:

TEMP.	OIL GRADE
below 0 degrees F.	5W-20
0 to 30 degrees F.	10W-30
above 30 degrees F.	30

Other manufacturers of air-cooled engines show the same reluctance to use multi-grade oils in the summer. BMW suggests 10W-50 below 86 degrees F and SAE 40 above that temperature or for substained high-speed operation. Two-cycles operate best on the specially formulated oils available from the major refineries or from the factory. In the case of ultra-high output engines, it is wise to follow the manufacturer's instructions as to the particular brand of oil.

Fuel

We think of gasoline as the fuel, but actually it only makes up a small part of the mixture. For every gallon of gasoline burned in the engine, the oxygen from 9,000 cubic feet of air is also burned. Normally, the air requirement is taken for granted, since most small engines operate where they have sufficient air supply. Probably the only exceptions are inboard marine engines and engines operated below decks on large vessels.

Gasoline that is produced from natural gas is called casing-gas, but most gasoline is processed from petroleum. To convert crude oil to gasoline, the liquid is heated in tall fractionator towers until the lighter compounds vaporize. These components are then cooled down to the liquid state and further processed. Primarily, gasoline consists of carbon and hydrogen atoms and various additives to increase the octane number, retard varnish formation, and to improve cold starting. The larger refineries match their product to seasonal changes.

The most important breakthrough in gasoline technology was made half a century ago by Charles F. Kettering. He discovered that trace amounts of liquid (tetraethyl) lead would significantly improve the anti-knock characteristics of gasoline. Previous to this discovery, gasoline engines were limited to compression ratios of three or four to one. Almost immediately, ratios went up to six or seven to one, and today many engines have ratios of eleven or more.

Without tetraethyl lead or an equivalent additive, gasoline has a tendency to detonate under compression. Detonation is a wild, uncontrolled explosion which occurs before the piston is at top dead center. Tremendous loads are imposed on the bearings and, on some engines, the operator can hear the distinct ping. In order to get some sort of standardization, the industry has agreed upon an octane number. The higher the number, the more "ping resistant" the fuel. In the United States, regular gas averages 94 octane and premium varies between 97 and 100 by the "research" method. Another way of calculating the octane number is the "engine" method which more nearly reflects commercial requirements. "Engine" octane is about ten points lower than "research" or passenger car octane numbers.

As long as an engine does not ping or knock (this will usually occur under load at part throttle) there is no reason to use higher octane fuel. At low rpm, carbon builds up in the chamber and on the top of the piston, increasing the compression ratio, and thus an engine might require higher octane as it ages. But this is unusual in small engines, which work hard and keep themselves fairly clean. High test will not give better economy than regular grades of fuel. All gasolines have about the same latent heat per pound. Nor is it necessary to run an occasional tank of high test through the engine to "clean" it.

Fuels with carburetor cleaners should not be used in two-cycles because these cleaners have a tendency to wash the oil off the parts.

Currently there is controversy about lead in gasoline. From a serviceman's point of view, low lead or no lead fuels are advantageous. With these fuels, there seems to be less spark plug fouling on two-cycles and there is definitely less oil contamination in four-cycles.

OPERATING PRINCIPLES

Conventional engines are classified as two- or four-cycle. Each upward or downward stroke of the piston is called a cycle: the two-cycle fires once every revolution of the crankshaft (or every two cycles), and the four-cycle fires every second revolution. As could be expected, each type has certain advantages. The two-cycle is mechanically simple; and in the smaller displacement classes, the two-cycle can generally be counted upon to outperform the equivalent four-cycle. But the cost of this performance is increased fuel consumption, poor part-throttle operation (although this problem is being surmounted), and, at least in some cases, excessive intake and exhaust noise. The two-cycle is favored in light-weight recreational equipment and in portable power tools. The four-cycle, on the other hand, suffers a built-in weight and cost penalty, which is compensated for by tractability, good fuel and oil economy, and ease of repair. At this writing, the four-cycle produces less air pollution than the two-cycle "smog motors." Some of the Japanese manufacturers promise a solution to this problem, apparently by the use of forced air injection at the exhaust port.

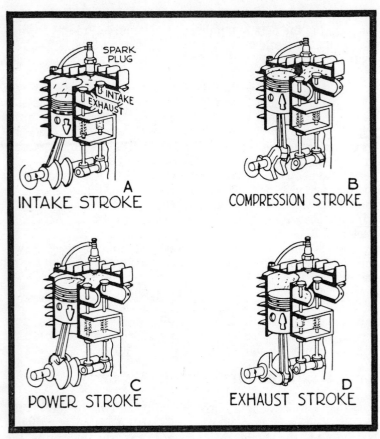

Fig. 1-6. Four-Cycle Operation. This is a side valve (L-Head) design; the basic operation is true for any four-cycle engine. (Courtesy Briggs and Stratton.)

Four-Cycle Operation

See Fig. 1-6. As the piston moves down on the intake stroke, the intake valve opens. Fuel and air enter the cylinder. The valve closes as the piston moves up, past BDC, and compresses the mixture on the compression stroke. Near TDC, ignition occurs, and the piston is driven down on the power stroke. The piston moves up on the exhaust stroke, expelling the burnt gases into the atmosphere through the open exhaust valve. At this point, the four cycles are complete and the process begins again with the intake stroke. The energy required to move the piston and work the valve gear is stored in the flywheel between power strokes.

The valves on a four-cycle operate from a camshaft which is driven at half engine speed. The cam is fitted with pear-shaped eccentrics, called lobes; tappets press against each lobe. The lobe-tappet arrangement converts rotary to reciprocating motion, which opens and closes the valves.

When the valves are positioned alongside of the piston, the engine is known as a flat or L-head, as in Fig. 1-6. This configuration is mechanically simple and, consequently, inexpensive to manufacture. However, the flat-head design is not really suitable for high compression ratios. These L-head engines are limited to applications such as lawnmowers and auxiliary power supplies. A variation of the side valve design is the T-head, which has a separate cam for the intake and exhaust valves. From a mechanic's point of view, this design is significant because both cams must be timed.

Overhead valve or I-head engines have the valves positioned in the head, directly over the piston as in Fig. 1-7. Such a configuration allows a compact combustion chamber

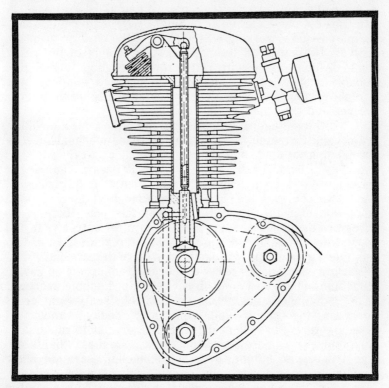

Fig. 1-7. Typical Overhead Valve Motorcycle Engine. Note the additional mechanical complexity when compared with L-head in Fig. 1-6.

29

and consequent high compression ratio. But this advantage is bought at the cost of additional parts; in order to transmit reciprocal motion from the tappets to the valves, the ohv design requires pushrods and bell cranks, called rocker arms. These extra parts add considerable inertia to the valve gear. While insignificant from the point of view of the power required to drive it, at high rpm the added inertia is a serious handicap. The combined weight of the tappets, pushrods, valves, and part of the rocker arms, overcomes the tension of the return springs and the valves stay open. The engine coasts, without compression, until the rpm drops enough for the springs to take hold. This phenomenon is known as "valve float."

Overhead-valve designs are found on a wide range of industrial engines (especially those which have come on the market through military surplus), and on a large number of motorcycles.

Gradually, however, ohv types are being replaced by overhead cams. The cam—or pair of cams—is positioned in the head, above the valves. Single overhead cams (ohc) employ fingers, working off the lobes of the cam, to open the valves. Since most of the reciprocating weight of the valve train is done away with, high rpm operation is possible. Double ohc designs go one step further and eliminate the weight of the fingers: one cam works directly on the intake valves and the other on the exhaust.

A disadvantage of the ohc is that the remote position of the cam calls for some fairly elaborate drive mechanism. The problem is not made any simpler by the fact that the cam has an uneven load. As the lobe revolves, it pushes against the resistance of the valve spring until the valve is fully opened. As the valve closes, the spring attempts to push the cam in the direction of rotation. At the heel of the lobe, there is no pressure against the valve, and the cam turns freely. In the past, a few English and Continental motorcycles have used a shaft and pinion gear arrangement to drive the camshaft.

Pure gear drives are out of the question, except on racing machines where cost is no object. The Ford double overhead cam Indianapolis engine used a total of twenty-eight gears between the crank and camshaft! Today almost all manufacturers use chain drive. Single or double row chains are found on Hondas, Triumphs, and Yamahas. Chains are quiet and cheap, but they do tend to stretch in operation, and a well-engineered design includes some sort of chain tensioner. Recently, there has been experimentation in the U.S. and in Europe on the use of toothed, fiberglass belts.

Two-Cycle Operation

The key to understanding the operation of these engines is to remember that the piston acts as valve and as a pump. As the piston moves up and down in the bore (Fig. 1-8), it uncovers ports which allow the fuel charge to enter and the

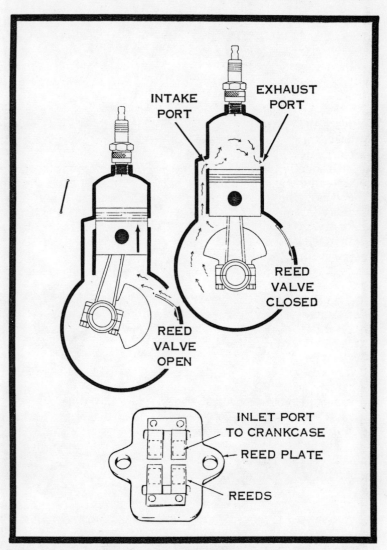

Fig. 1-8. Typical Two-Cycle Engine with Reed Valve Assembly. Some high-compression engines may have 2 reed valve plates, each fed by its own carburetor. (Courtesy Outboard Marine Corp.)

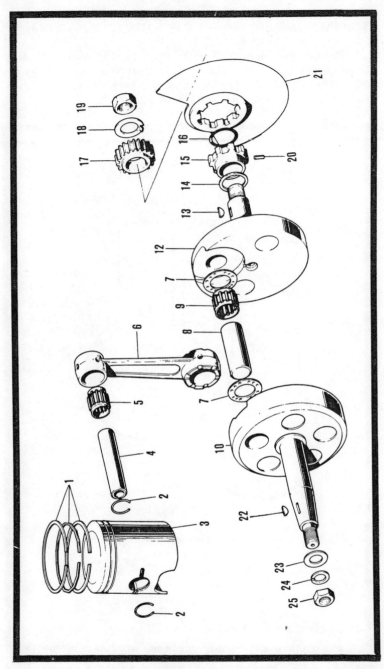

Fig. 1-9. Disc valve used on Kawasaki Two-Cycle Motorcycle. (Courtesy Kawasaki Heavy Industries, Ltd.)

exhaust gases to escape. At the same time, the underside of the piston functions as a kind of pump.

When the piston moves up in the bore on the compression stroke, it leaves a low pressure area behind it. Fuel and air move into the crankcase to fill this partial vacuum. On the power stroke, the piston moves down in the bore, pressurizing the crankcase, and forcing the fuel charge into the bore by means of a transfer port which empties into the bore. Of course, the crankcase is air tight and there must be some method of preventing reverse flow through the carburetor. American designs, such as West Bend and Power Products, often employ a shutter or reed valve between the crankcase and the carburetor (inset, Fig. 1-8). When there is low pressure in the case, the spring steel reeds open and fuel passes through the valve. Under compression, the reeds close against their backing plates, trapping the mixture in the case where it will be pumped through the transfer port.

Kawasaki, Bridgestone, and a few other makes, use a fiber disc rather than reed valves. The disc has a cutaway on the rim, and the disc is timed to piston motion (Fig. 1-9). A variation of the disc valve principle is found on model airplane engines; the crankshaft has a passage milled along its length and opening to the crankcase. At the moment of maximum crankcase vacuum, the passage opens to the carburetor, and fuel flows. Some engines are entirely piston-controlled (Fig. 1-10).

So far we have discussed events below the piston. But that is only half of the story. As the piston moves down the bore, it uncovers one or more exhaust ports. Exhaust gases, which are at greater than atmospheric pressure, escape out of the port as soon as it is uncovered. However, a more positive method is needed to insure complete evacuation, that is, scavenging. The fuel charge, entering the chamber through the transfer port, is used to force the exhaust residues out through the tail pipe. Of course, some of the fuel will mix with the exhaust and be lost. The older designs depended upon the shape of the piston crown to deflect the incoming charge away from the exhaust. A deflector piston is pictured in Fig. 1-11. While adequate for some applications, deflector pistons are heavy and prone to over-heating. Some of the newer designs employ light-weight, flat-topped pistons. The charge is introduced through multiple ports which are angled to give the mixture a swirl. In all of this discussion, it should be remembered that the gas-air vapor induction speeds are high—in the range of 900 feet per second. A deflector or a slight angle on the ports will have tremendous effect.

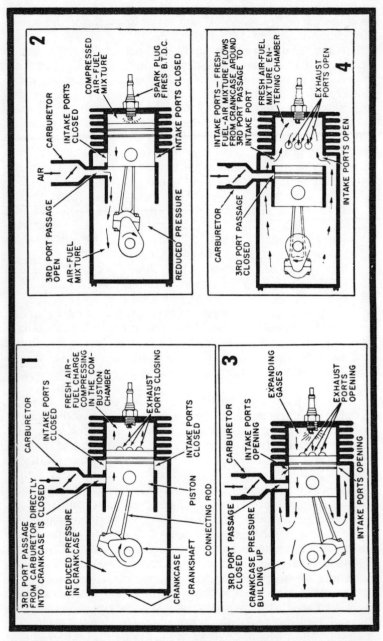

Fig. 1-10. Operation of Piston-Controlled Induction System. Note that piston is flat-topped, and scavenging depends upon the swirl imparted to the mixture entering the combustion chamber. (Courtesy Tecumseh Products.)

OVERALL TROUBLESHOOTING PRINCIPLES

The sections that follow cover overall troubleshooting techniques for isolation of the system that is faulty. Once you've established the faulty system, you should refer to the chapter on that area, where detailed troubleshooting and repair are covered.

Ignition Checks

According to one study, nearly 9 out of 10 small engine failures are caused by a breakdown in the ignition system. Many experienced mechanics assume that the ignition is at fault even when the evidence seems to point to the fuel system or to weak compression. And most of the time the mechanic is right.

Begin by looking for the obvious. If the engine has a switch, see that it is on. If equipped with a battery, check that it is charged. On motorcycles, this is easily done by sounding the horn. Another method is to short the negative and positive terminals momentarily with a pair of pliers. There should be a heavy splatter of sparks.

Disconnect the spark plug wire or high tension lead, as it is sometimes called, and hold it so that the end is about 3-16" from the block. Fig. 1-12 shows how a spark plug adapter and alligator clip can be used in a safe manner. Being careful not to touch the metal terminal, crank the engine. You should hear a "snap" and see a heavy, blue spark jump from the terminal. A thin spark or one that is any other color than blue means that the ignition is defective.

Assuming that the spark is OK, the next step is to check the spark plug. The porcelain insulator should be clean and unbroken. A crack in the insulator will short circuit the plug, rendering it useless. The nose of the plug—the part that fits into the engine—should be reasonably clean and dry. While the plug is out, position the metal part against the engine, connect the high tension lead, and with your hands away from the ignition parts, crank. It should fire; if it does not, a new plug is clearly indicated. But a word of caution: many times a spark plug will give off a healthy blue spark in the atmosphere, and refuse to fire in the engine when it is working against compression. The only valid test is to substitute a known good plug, i.e., one that runs in a similar engine. Even brand new plugs are not to be entirely trusted. Many a poor mechanic has been driven up the wall by that one new plug out of a hundred or so that is faulty.

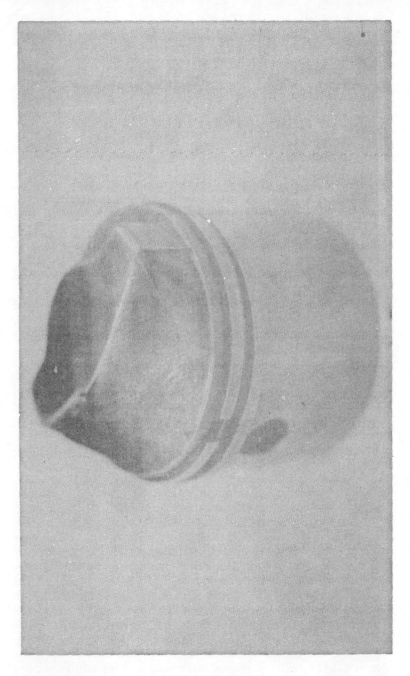

Fig. 1-11. Deflector Piston from a Power Products Engine. (Photo by Al Galinski.)

Fig. 1-12. Using a Test Plug to Check Ignition on a Homelite Chain Saw. The spark should be heavy and blue. (Courtesy Homelite, a Textron Division.)

Fig. 1-13. Check the Choke Position. Choke plate "E" must be closed completely for cold starting. (Courtesy Kohler of Kohler.)

Fuel System Checks

Again, begin with the obvious. Is there gasoline in the tank? Is it clean? Often you will find that water has condensed in the tank. And gasoline does deteriorate with age. The highly flammable aromatics evaporate, leaving varnishes and oils which are difficult to ignite. If you find contaminated fuel in the engine, chances are that the owner also has gasoline on hand which is also contaminated in the container. Is the fuel tap on? A few motorcycles have a vacuum valve on the fuel line which can fail.

See if gasoline is actually reaching the carburetor. Walbro carbs (fitted to Tecumseh and Clinton engines) have a float chamber drain for this purpose; when depressed, gasoline should flow out and around it. English Amal and a number of other motorcycle carbs have a "tickler" positioned above the float. It acts to flood the carburetor. Many outboards have a brass drain plug which is screwed into the side of the chamber. Other types must be partially disassembled to determine if they are receiving fuel.

Power Products diaphragm-type carbs, which are becoming nearly universal on lawnmowers, chain saws, and go-karts, require a special procedure. First, remove the air filter. Then locate the small hole drilled into the bottom of the diaphragm cover. Insert a piece of stiff wire or a small Allen wrench into the hole and push upward about $\frac{1}{8}$". After a few seconds, gasoline should dribble out of the air intake.

Suction-type carbs, sometimes called "mixing valves," are found on many Briggs and Stratton engines and on some of the inexpensive Sears, Roebuck lawnmowers. Identification is easy: this type of carb is always mounted directly on top of the gasoline tank. They have no float, though some have diaphragms, and work like a flit gun—fuel is sucked through a tube from the tank, and then metered through an adjustable jet and into the engine. These devices are very reliable, if fuel is not allowed to stand in the tank for a long period. When this happens, the pick-up tube becomes clogged and the engine will not start.

Now is a good time to inspect the choke. Is it closing completely? On many lawnmowers with remote controls, the choke linkage can work itself out of adjustment and cause hard starting (Fig. 1-13).

Paper air filters can also cause hard starting. Most modern engines are equipped with them because paper is the most efficient filter element known. The problem is moisture. A wet filter will swell and prevent air from entering the engine. Most filters have been engineered so that they will stay dry even in a driving rain—but an over enthusiastic application of a water hose or an attempt to clean one of these filters with solvent will ruin it. The only fix is to throw away the soaked filter, dry out the element holder, and replace the element.

Having made the above checks, crank the engine through a few times, stop, and remove the spark plug. The nose of the plug should be just damp and smell of gasoline. Too much gas is as bad as too little. If the plug is soaking wet, suspect that the filter is stopped up or (more likely) that the engine has been cranked too much with the choke full on. The remedy is to rapidly spin the engine with the plug still out. The up and down movement of the piston will cause a draft which will eventually dry the chamber. Another, faster, way is to blow the chamber out with compressed air, but be sure your compressed air supply is clean and dry.

There are two special cases concerning flooding. Two-cycle engines can fall victim to crankcase flooding, that is, the crankcase can become partially or completely full of liquid fuel. Some engines are provided with a crankcase drain just for this eventuality. However, the usual procedure is to crank until the fuel is forced out of the crankcase and into the combustion chamber, where it will eventually evaporate. The other special case is oil flooding. This can occur in a two-cycle if the fuel mix is too "rich" in oil (12 parts gasoline to 1 part oil is the absolute limit—most engines operate best at between 16 and 20 to 1 and a few outboards can go as lean as 100 to 1 with the proper oil). An oil bath air filter which has been overfilled can also cause oil to get into the combustion chamber. The most serious oil flooding occurs in four-cycles which have been tilted so that the head is down. Oil runs out of the crankcase, down the bore, past the piston rings, and into the chamber. The standard procedure is to first flush the chamber with lacquer thinner or lighter fluid, let it dry, and try again.

Compression Checks

A rough and ready way to check compression is to remove the spark plug and hold your thumb over the hole. As the engine is pulled through, you should feel an alternate blowing and sucking action. On two-cycles this will occur with every revolution, on four-cycles with every second one. But a compression gauge is more accurate and quite inexpensive.

When making the compression test, you should listen closely for unusual noises. Scraping or squealing (Fig. 1-14) means that a bearing surface has failed. A tapping sound once every revolution means a very loose connecting rod. For either problem, teardown is indicated.

Suppose that the engine develops no compression. On four-cycles, this most often means that a valve has stuck open. You can usually free the valve without removing the head. On two-cycles, the problem is usually a burnt piston. But on either type, the complete absence of compression can mean a thrown rod.

Weak compression—and here is where experience or a gauge is necessary—can be caused by a number of factors. In general, two-cycle engines are low-compression devices to begin with. Many of them will not give a reading of more than 60 pounds per square inch at cranking speed, while equivalent four-cycles will be in the range of 110 to 140 psi. Some four-cycles have a compression release which automatically comes into play during cranking. The most common of these is the "Easy Spin" system on late-model Briggs and Stratton products. To get an accurate reading on these engines, the flywheel must be spun counterclockwise.

The usual causes of low compression are leaking valves, worn or stuck rings, and leaking head gaskets. Compression leaking past a head gasket can often be heard as the engine is

Fig. 1-14. When checking for compression, listen for unusual noises. (Courtesy Outboard Marine Corp.)

spun over. Some two-cycles used on motorcycles have a manually-operated compression release operated from the handlebar. Sometimes a release—which is no more than a tiny poppet valve—fails to seat. You can feel the air escaping around it.

TROUBLESHOOTING SPECIFIC SYMPTOMS

Failure To Start When Hot

Perform the same general checks previously described. Work quickly though, before the engine cools. If the ignition breaks down, suspect the coil and the condenser in that order. After you have tested the ignition, look for flooding. This condition can be caused by a choke butterfly which is not opening completely, an over-rich adjustment on the main jet, or by a float level which is too high. Compression in this case should give no trouble—if an engine has enough compression to start when cold, it should start hot.

Before you return the engine to regular service, run it for an extended period and repeatedly check for hot starting. An engine in good tune should start on the first or second attempt, whether cold or hot.

Lack of Power

Ascertain for yourself whether or not the engine is performing as it should. Many times, people have an exaggerated notion of the potential of their machine. While you're at it, make a thorough examination of the running gear. On lawn-mowers, see that the blade is sharp and that the propulsion mechanism (if fitted) is working correctly. Often "lack of power" is no more than a loose belt. On motorcycles, look for a slipping clutch, under-inflated tires, over-tight chains, and dragging brakes. Check the track and drive system on snowmobiles.

But if all else is OK, then we go to the engine, where "lack of power" can mean two separate problems. One, the engine could be missing at high speed. This is usually the fault of the spark plug or of some other component in the ignition system. Or, missing can be caused by a seriously misadjusted main jet, or even by a clogged air filter. The second case is where the engine seems to be running smoothly enough, but still fails to produce power. Suspect the timing first. Then adjust the carburetor slightly richer. If the problem persists, run a compression test. On two-cycles—especially those which are operated at moderate speeds—you will often find the exhaust ports partially clogged with carbon.

Erratic Idle

First check the timing and then the point gap. Next, adjust the low speed needle on the carb. If the engine still won't idle, you can remove the needle and attempt to blow out the idle circuit with compressed air regulated to 30 psi. But a word of caution—blowing on the float chamber vent can cause the float to collapse. Many outboards have a diaphragm type fuel pump which operates from crankcase pressure. If the diaphragm is ruptured, rough idle will result. Check for air leaks in the intake tract—at the joint between the carb and intake manifold, and between the manifold and the head. You can do this check quickly by squirting very small amounts of gasoline on these joints. If the engine speeds up, there is a leak. Look for signs of wear on the throttle shaft. On a two-cycle, erratic idle can be caused by leaking crankcase seals. And finally, on industrial engines, check the governor linkage for excessive play.

Overheating

How is the engine being used? Heavy overloads will inevitably cause overheating. And motorcycles (with the exception of a few fan-cooled models such as the small Ducati) are designed on the assumption that they will be moving. Long periods of stationary operation can destroy one.

If the problem is real, then begin by checking the lubrication system. See that the oil level is correct and that the proper grade of oil is being used. Next, check that the spark plug is the proper heat range as recommended by the manufacturer for that particular engine. Make sure that the carburetor is adjusted properly and that there are no leaks in the intake tract. Check the ignition timing.

The cooling system can also be at fault. On outboards, see that the pump is putting out sufficient water. On engines with underwater discharges, run the engine a few minutes; stop it and hold your hand against the cylinder. You will feel the cylinder grow steadily warmer over a two or three minute period. Now start the engine—if the pump is working properly, the head will begin to cool almost immediately. On air-cooled engines, inspect the fins to see that they are free of accumulated oil and dirt (Fig. 1-15). If the engine is equipped with shrouding, make sure that it is intact and installed correctly.

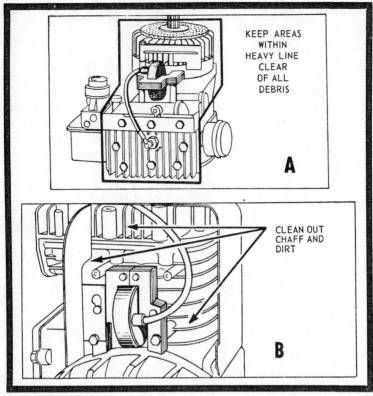

Fig. 1-15. Keep these areas of Air-Cooled Engines clean or severe overheating will result. (Courtesy Briggs and Stratton.)

Dunking (Immersion in Water)

Outboard engines occasionally "go down," that is, are accidentally and totally immersed in water. The smaller engines should be equipped with safety chains for easier retrieval, but safety chains on the larger units can pull the transom out. To determine what course of action to take, it is necessary to know exactly what happened. If the engine rested on a soft bottom, you can assume that it is silted and must be torn down completely. On the other hand, if it went down in clean water, it may be possible to restart it immediately without damage. But a running engine, especially one that was at or near full throttle, might have damaged itself by ingesting water. Time is also an important variable. Immediate "first aid" can save a tear-down. Even a few hours in the air with the unit inoperative can start a corrosion problem which can only be corrected by full disassembly.

To re-start a dunked engine, it is necessary to work quickly. Every minute the unit is exposed to air counts. Remove the cowling, flywheel, carburetor and reed valve assembly, and the spark plugs. Turn the engine so that water will drain out of the crankcase. If methylal (methyl alcohol, NOT rubbing alcohol) is available, splash it around the crankcase to absorb the water. Crank the engine repeatedly. The magneto coils should be dried and the points wiped off. Disassemble the carburetors and fuel pump, and drain the gasoline tank. It is wise to spray all exposed metal surfaces with a preservative such as WD-40. Assemble and run the engine for at least 30 minutes to evaporate any standing water. Later, if the engine begins to knock, expect that the lower main bearing was not completely dry. This is the most vulnerable friction surface on the engine and even a tiny speck of rust will ruin the bearing.

Racing mechanics can re-start a dunked engine in five minutes or less. Their procedure is to drain the carburetor float bowls, wipe the points dry, and drain the water out of the cylinders through the spark plug holes. If the engine starts, it is assumed that it will be OK until the next overhaul.

Racing outboards have cut-down flywheels which minimize hydrostatic damage. Hydrostatic lock occurs when an incompressible liquid—in this case, water—is trapped between the top of the piston and the cylinder head. The inertia of the flywheel and the crankshaft slams the piston against the column of water. Something has to give. It may be the head which cracks, a sheared flywheel key, a rod, or the crankshaft. There is also the very real possibility of warpage of the cylinder bores and the pistons, caused by the sudden cooling of these parts. Obviously, engines with such kinds of damage must be torn down for repair.

The important thing is to prevent corrosion during the interim. The entire unit, or the power head on larger engines, should be immersed in a tub of oil or fresh water. If fresh water is not available, it is better to use salt water than to allow the unit to be exposed to air.

As soon as you can, start the teardown. Delaying repair more than 12 hours is asking for trouble. But if delay is unavoidable, say over a week-end, the engine may survive if kept immersed. Expect, though, that the coils might become shorted. Dry them under a lamp, never in an oven. The powerhead should be torn down completely. Check the crankshaft, rods, and the cylinder bores for trueness. Each part should be washed by hand and immediately oiled to remove any trace of water, salt, or silt. Electrical connections should be scraped bright, and the cooling passages should be thoroughly flushed with kerosene to clear them of debris, and then bake them dry under a lamp. It is good practice to paint the engine to prevent corrosion formation on its other surfaces. Use a paint especially formulated for adhering to aluminum, but avoid marine anti-fouling paint. These paints contain metals which will increase corrosion on aluminum.

Chapter 2

Ignition Systems

A few years ago, one American small engine manufacturer discovered that 87 percent of customer complaints involved the ignition. Some mechanics of many years experience will automatically assume that the ignition is bad. This attitude probably saves time, especially where flywheel magnetos are concerned.

Anyone can understand how an engine or a transmission works by studying the relationship between the parts. But this does not hold true of a magneto or a generator. In order to understand an electrical machine, you must have some knowledge of the theory.

An analogy will go far to explain the nature of electricity: think of an aluminum pipe filled with ping-pong balls. When a ball is added at one end, all the balls move down the pipe and the last one drops out, where it can be retrieved and put back into the pipe.

The balls represent electrons—tiny bits of charged energy—and the pipe is the conductor. The sides of the pipe act as insulation by preventing the balls from escaping before they reach the terminal end. Voltage is the force or pressure exerted on each ball as it is pushed into the pipe. Amperage represents the number of balls which drop out over a unit of time. And finally, the pipe, as does any conductor, offers a certain resistance to the movement of the balls.

Electrons flow from the negative terminal (called a pole) of a battery or generator and make a complete circuit back to the positive terminal. Think of the pipe—if we could not retrieve the balls, there would be no current flow. An open circuit means that the path of electrons is blocked; a short is equivalent to a hole in the side of the pipe. Rather than overcome the resistance along the length of the pipe, the electrons take the easy way out and escape through the hole.

Some materials are better conductors than others. Silver, lead, copper, aluminum, and carbon make good "pipes," while air, rubber, plastic, and wood resist the flow of electrons, and therefore are insulators. But these terms are relative; all conductors have a certain resistance relative to the material, the cross-section and the length, and all insulators can be overcome with a high enough voltage. To ignite the fuel-air mixture in a combustion chamber, electrons must jump an air gap at the tip of the spark plug. All it requires is sufficient pressure, in this case, about 16,000 volts.

Amperage, or the amount of current, is also important. Voltage alone cannot do work. A 6-volt lantern battery will not crank an automobile, simply because it does not provide enough electrons for the starter motor. Or take an ordinary black and white television set: the high voltage side, going to the picture tube, has a potential of about 20,000 volts, and very little amperage. If you were to touch the terminal, you would be hurt, though hopefully, not electrocuted. The low voltage side is another story. It may produce only 6000 volts or so, but have enough amperage to be deadly. Common house current at 220v and 30 or more amperes can kill.

Small engines (and automobiles) use a single wire system. The engine block is part of the circuit; one side of the battery is grounded to the block and the other side, the "hot" side, carries current to the various components which are themselves grounded. This method saves wire, but frayed insulation on the hot side can easily develop into a short circuit. For this reason naval vessels and aircraft commonly use the two wire system. Before shorting can occur, uninsulated portions of both conductors must be in contact.

Electricity and magnetism are closely related, but distinct. Whenever an electric current flows, a magnetic field is generated around the conductor. This phenomenon accounts for the erratic behavior of a compass in the neighborhood of electric power lines. Nothing known to man can shield magnetic force—it penetrates everywhere. But soft iron, while it cannot block the force, has the mysterious ability to draw and focus it. Wrist watches which are labeled "anti-magnetic" merely are encased in a covering of soft iron. Rather than go through the workings of the watch, the magnetic lines of force concentrate and pass through the iron cover. Coils, solenoids, and motor armatures, which depend upon concentrated magnetic force, always have a core of soft iron.

Electricity produces magnetism, and magnetism can be used to produce electricity. If you were to take a piece of copper wire (or any other conductor) and pass it between the poles of a magnet, a small current would be induced in the wire. The same effect would be produced if you held the wire stationary and moved the magnet. In either case, a magnetic field moving relative to a conductor generates current in that conductor.

Coils

The heart of the ignition system, the coil (or transformer), works on the principles discussed in the preceding two paragraphs.

Coil Construction

As illustrated in Fig. 2-1, an ignition coil consists of two windings, one on top of the other, wound around a soft iron core. The primary, or low voltage, winding is made up of approximately 200 turns of heavy copper wire. The high voltage, or secondary, winding consists of 10,000 or more turns of extremely fine wire. One end of the high voltage lead plugs into the coil and the other end connects to the spark plug, either directly or, on some multi-cylinder engines, via the distributor. The primary is either connected to the battery or is so situated in a magnetic field that it becomes charged. The primary is wired to the points so that current flow is interrupted as the crankshaft turns. When the points close, heavy current flows through the primary, saturating the secondary with magnetic force. As soon as the points open, the current flow abruptly ceases. The magnetic field in the primary side of the coil drops from full strength to zero. This moving (collapsing) field induces a high voltage in the secondary which fires the plug. The voltage increase in a coil is not getting something for nothing, although it might seem like it. As the voltage is increased, the amperage is reduced, so additional energy is not created. The current is transformed. The greater the ratio between the number of secondary and the number of primary turns, the greater will be the transformation. A coil with a ratio of 400:1 should produce twice the voltage—but only half the amperage—of one with a ratio of 200:1. Fortunately, it takes very little amperage to ignite the fuel-air mixture.

Coil problems are difficult to diagnose. This is the reason why the coil is the last item in the ignition circuit to be tested—if everything else checks out OK, the mechanic can assume that the coil must be grounded to the engine or the frame, since the high tension side is internally grounded. A loose coil or one that is bolted to a heavily painted surface will not function properly. Also, coils are wound with a particular polarity. It is possible to connect the type shown in Fig. 2-1 backwards and so reduce its efficiency.

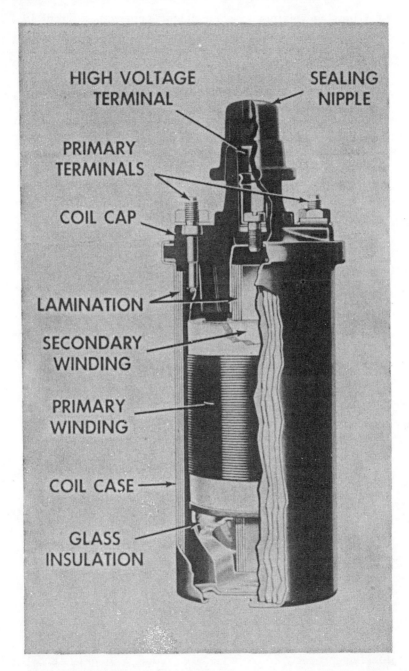

Fig. 2-1. Typical Ignition Coil used in Battery systems. (Photo courtesy Delco-Remy.)

Coil Polarity

Commercial polarity testers are expensive and even the kit model sold by Knight costs nearly $100. An inexpensive way to test coil polarity is to use an ordinary lead pencil. Wrap the metal eraser end to reduce the possibility of shock, and start the engine. With the engine running, remove a spark plug wire and hold it within a half inch of the plug so that the spark will jump from the terminal to the plug. Insert the sharpened point of the pencil in the path of the spark. A coil with correct polarity will cause the spark to flare on the spark plug side; incorrect polarity will show itself by flaring on the high tension wire side of the pencil. On engines with negative grounds (as defined by which battery post is grounded to the frame or the block), the hot wire from the ignition switch should be on the positive terminal of the coil. Most coils have this terminal marked with an "I." The negative or minus terminal of the coil should be connected to the distributor or the points. On positive-grounded systems, the wire from the switch goes to the negative terminal of the coil and the breaker point lead connects to the positive terminal.

Coils may suffer from accumulated corrosion at the terminals. Both the primary and the secondary terminals should be scraped clean and checked for tightness. Corrosion in the primary circuit means resistance and lessened current flow, which in turn means a weaker magnetic field because the field depends upon the amount of electrons flowing. The secondary lead is very susceptible to corrosion. Usually, it plugs into the top of the coil, although on Lucas designs it is fastened into place with a threaded cap. On Briggs and Stratton coils the secondary is merely twisted through a terminal. This is adequate; over-zealous mechanics have overheated and ruined these coils by soldering the wire in place.

Coil Testing

There are three methods of testing a coil. One is to use a commercially available tester such as the Merc-O-Tronic unit. These testers have a scale for resistance checking of both sides of the coil (low resistance would mean that some of the windings have shorted together, and high resistance would mean a break in the windings). They can also test power. The primary is energized with intermittent direct current (DC) and the secondary output is measured. Each coil has its own test specifications, which vary to some extent with the make of tester used.

Fig. 2-2. Four most-common spark plug gap configurations. (Photo courtesy AC Spark Plug Div., General Motors.)

Another method is to connect a battery to the primary side of the coil. (On battery ignition systems, the machine's own battery can be used. Coils for magneto systems will require 12 to 18 volts which may be obtained by connecting batteries in series.) Rapidly make and break one connection and observe the high tension. With each make and break there should be a spark from the end of the lead to the ground. This test is not infallible, and should not be used if a good coil tester is available. The third method is to substitute a known good coil for the suspected one. Most mechanics prefer this method since they work from an inventory and have the parts on hand.

Different types and makes of coils can be substituted. This is especially true of battery ignition systems. On magneto systems, the choice of coil is much more critical and there is the problem of the size of the laminations. On battery systems, two factors are important: the voltage, since 6 or 12 volts do not interchange; and whether or not the coil has an internal resistor. Some coils have a built-in resistance to protect the points from heavy amperage. If a coil without resistance is put into a system which was designed for one, an external resistor should be connected into the primary circuit. However, this consideration is limited, in general, to marine engines. Most small engines do not use any resistance in the primary.

SPARK PLUGS

The most abused part of the ignition system is the spark plug. Its outer portion is more or less exposed to the elements and to accident, while its firing tip is expected to operate in temperatures which approach 2,000 degrees F. The first part that a mechanic will inspect in a baulky engine is the spark plug.

There are many styles of spark plugs, as shown in Fig. 2-2. Reading from left to right in the photo, the first type is rarely seen except at race tracks; this pin gap type is used in ultra-high compression engines where there's very little room in the combustion chamber. The second from left is the plated plug, surface gap type; there isn't an air gap—the spark jumps from the insulator to the grounded shell. The second type is found in two-cycle engines with capacitance discharge (CD) ignition. The third plug is the clipped or J-gap, used in many small engines. Because the side electrode extends only halfway across the center electrode, this plug is less liable to foul. The fourth plug, of course, is a standard automotive type.

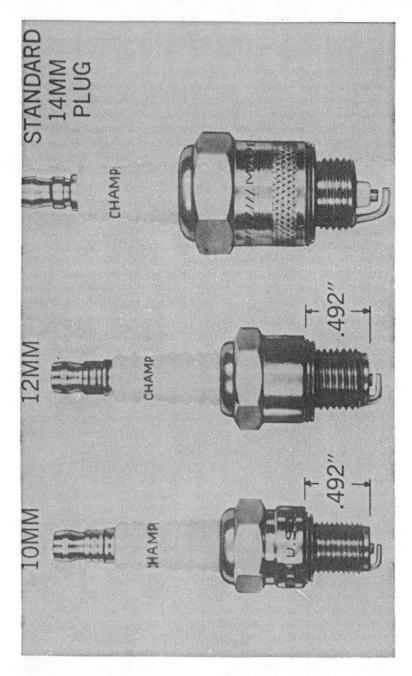

Fig. 2-3. Spark Plug Reach vs Bore Dimensions. (Photo courtesy Champion Spark Plug Co.)

Spark Plug Construction

All spark plugs are built around three basic elements. The metal part—the shell—holds the plug into the engine, provides a heat path away from the firing tip, and serves as an electrical ground. The insulator, still called the porcelain because the first ones were made of that material, insulates the center electrode from the shell. The center electrode is a metal rod which carries voltage from the coil. In certain applications, it may have a built-in resistor to reduce radio interference.

The standard thread size is 14 millimeters. Smaller, 10 and 12 mm plugs are found in motorcycles where space is at a premium. Occasionally you will encounter the 18 mm Ford type or even the ancient ¾" "pipe thread." While not the best solution, it is possible to salvage a stripped cylinder head by tapping it out to the next size. Be sure though, that the substitute plug matches the original at least in tip length and heat range, items discussed below.

In addition to different bore diameters, spark plug shells are made with different reaches (Fig. 2-3). The reach is the distance measured from the seat to the end of the threads and varies from ⅜" to ¾". Each engine is designed for a plug with a specific reach: substitution of a shorter one can cause difficult starting and a longer one can result in spark plug-piston collision.

Heat range is another variable. Fig. 2-4 illustrates the principle. Heat travels away from the tip, through the insulator, and out to the shell. The further the heat must travel, the hotter the tip will be. Different engines demand spark plugs of different heat ranges. Some run relatively cool, and so require a hot plug to insure good combustion; others, especially the high-revving, air-cooled types, need a cooler plug in order to prevent pre-ignition. On American plugs, the heat range increases with the type number. Thus, an AC 46 is one step hotter than an AC 45. The European practice is just the reverse—the hotter the plug, the lower the number. In any case, small engine mechanics rarely deviate from the manufacturer's specifications. However, if an engine continually fouls plugs because of excessive oil consumption, it is safe to go one step hotter. If, on the other hand, the engine burns the recommended plugs because of over-revving or other abuse, the mechanic can substitute a cooler plug.

Spark Plug Tests

It is possible for an experienced mechanic to tune an engine by "reading" the plugs. First, make certain that the

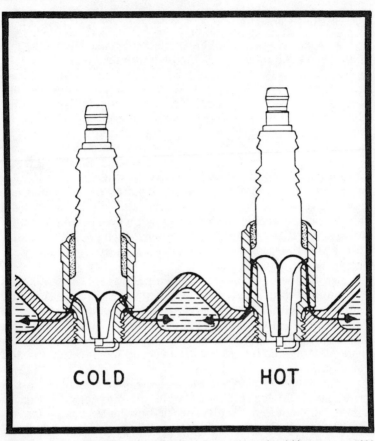

Fig. 2-4. Spark Plug Tips. Ideally, tip temperature should be approx. 1000 degrees F. Tip temperature is regulated by insulator length. The further the heat must travel, the hotter the plug. (Photo courtesy Champion Spark Plug Co.)

plugs are those recommended by the manufacturer. Then get the engine up to normal temperature and operate for a few minutes under load. Without closing the throttle, cut the ignition and inspect the firing tip. A light brown tip, say, the color of coffee with a dash of cream, means that combustion temperatures are normal. A black, sooty tip means an over-rich mixture, and a white tip means that the mixture is lean. This latter condition must be corrected immediately in order to prevent piston and valve damage. The following detailed information on reading plugs refers to Fig. 2-5:

(1) Normal, insulator light brown, engine in good tune and spark plug is correct heat range.

(2) Worn out, the electrode is thin and rounded.

(3) Oil fouled, the insulator is covered with wet, sticky oil deposits. On four-cycles, check the condition of the rings, valve guides, crankcase vent, oil level in crankcase and in oil bath air cleaner. On two-cycles, check the fuel-oil mix for the correct proportion.

(4) Splash fouled, merely shows that the engine has responded to a recent tune-up. The condition will disappear in time.

(5) Core bridging, carbon build-up between the insulator and the shell. This is caused by the same conditions as described in (3), or may be the result of excessive idling.

(6) Gap bridging, deposits between the center and ground electrodes. This condition is very common with two-cycles, and seems to occur at random intervals.

(7) High speed glaze; hard, glassy deposits on insulator. The glaze is conductive and will ruin plugs. The problem can be traced to long periods of idling combined with sudden acceleration.

(8) Scavenger deposits, brown, yellow, or white crystal-like deposits on the insulator and side electrode. It is caused by some gasolines or by extended low speed operation.

(9) Aluminum throw-off, molten pieces of aluminum on the plug. This condition is serious and calls for an immediate tear-down to inspect the piston. It is usually caused by ignition timing which is too far advanced.

(10) Preignition, insulator blistered, white from overheating. Check the timing and use a better grade of gasoline.

Spark Plug Maintenance

Maintenance is largely a matter of periodic cleaning and gapping. The best way to clean plugs is with a sand blaster. (Because auto mechanics no longer bother to clean plugs, many of these machines are on the second-hand market and can be purchased for as little as five or ten dollars.) Without a sand blaster, plugs can be cleaned by soaking in paint remover, or, in an emergency, with a wire brush. After cleaning, the inner surface of the two electrodes should be dressed with a point file. Flat, sharp-cornered surfaces make it easier for the arc to jump.

The gap will vary with different engines, but this information is usually stamped on the magneto or on the specification plate. A good average gap figure is .025". Many mechanics prefer to use a wire-type gauge, but the familiar, flat-bladed gauge is just as accurate. The ground electrode—not the center one—is bent until there is a barely perceptible drag as the gauge is passed between the electrodes.

3—OIL FOULED

5—CORE BRIDGING

2—WORN OUT

4—SPLASH FOULED

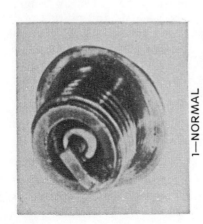
1—NORMAL

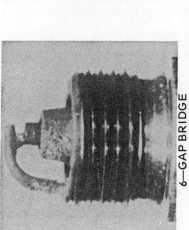
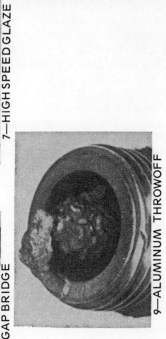

6—GAP BRIDGE 7—HIGH SPEED GLAZE 8—SCAVENGER DEPOSITS 9—ALUMINUM THROWOFF 10—PREIGNITION

Fig. 2-5. Troubleshooting Engine Condition by "Reading" the Spark Plug.

To install, start the plugs by hand, and torque to the following specifications:

TORQUE
(Foot-Pounds)

Plug Size	Aluminum	Cast Iron
10 MM	9-11	13-15
12 MM	15-20	20-35
14 MM	18-22	26-30
14 MM - Tapered	10-15	10-15
18 MM	28-34	32-38
18 MM - Tapered	15-20	15-25
7/8" 18	31-39	35-43

HIGH TENSION CABLES

Sometimes called the high tension leads, these are the wires from the coil to the spark plugs. They carry a minimum of 16,000 volts, and, in the newer capacitance discharge systems, they may carry as much as 50,000 volts. It is obvious that with this kind of voltage, the insulation must be almost perfect. Two basic types are available: the radio suppression and the metallic core. By Federal regulation, all motor vehicles used on the public roads must be equipped with radio suppression cables. The ignition circuit is really a kind of broad wave radio generator which broadcasts each time the points open. The "program" is picked up in the form of static on nearby radios and television receivers. When you see a series of dashes march across your TV screen, you can be sure someone is driving by in a car equipped with old type metallic-core cables.

High Tension Cable Construction

Radio suppression cables have a graphite core. On the earlier types, granules of graphite were packed into a tiny plastic tube and covered with insulation. As the engine ran, vibration caused the graphite to settle at the lowest point on the cable, leaving gaps in the path of the spark. Bad experiences with this early type have caused mechanics to overreact and condemn the whole tribe. Today's radio suppression cables are quite reliable—if handled correctly.

Other types of cable have a core which consists of braided nylon saturated with particles of carbon or graphite. These cables are strong, and resist the effects of rough handling, but they suffer from poor connections to the terminals. Eventually the core will pull free and the spark will begin to arc, burning away the core, the insulation, and, in extreme cases, the coil terminal. Never pull or stretch these cables. Remove them by pulling on the terminal. Be especially careful in cold weather, as the nylon core becomes brittle and subject to cracking.

The best cable now available has a tubular core which is resistant to shock and vibration. Belden No. 7300 cables can actually be tied in knots without damage.

Testing of Cables

The test for these cables is the amount of resistance between terminals as measured by an ohmmeter. Some resistance is designed into the circuit (this figure varies with different manufacturers between 7,000 and 15,000 ohms per linear foot). A defective cable will usually betray itself by astronomical resistance readings, in the range of 150,000 ohms per foot to infinity.

Metallic-core cables are used in stationary power plants where radio suppression is of little importance. These cables should have only 1 or 2 ohms per foot resistance. Three kinds of material are used in the core: steel which is durable, but which can cause serious corrosion problems in wet climates, copper, and silver-plated copper. The latter has by far the best electrical properties and is available as Packard 440.

Both radio suppression and metallic-core cables can be purchased with various types of insulation. The most common type consists of a rubber insulator encased in a neoprene jacket. The neoprene is resistant to chemical attack, but is easily damaged by heat. A better insulation is silicone plastic such as found on Belden No. 7766 and Autolite 7SH. The major disadvantage of cable insulated with silicone is that it is expensive, costing a dollar or more a foot.

Metallic cable is usually purchased in bulk and cut to fit. Make the leads as short as possible but make sure they don't contact hot or vibrating parts. On magneto engines, excessive slack in the cable can cause contact with the flywheel. Grommets should be used wherever the cable is routed through a metal part such as a loom or a shroud. On multi-cylinder engines, keep the cables as far apart as possible in order to preclude "cross-firing." Although few mechanics take the time to do it, terminals should be carefully soldered with ROSIN-CORE solder so that the wire and the terminal make a zero-resistance connection.

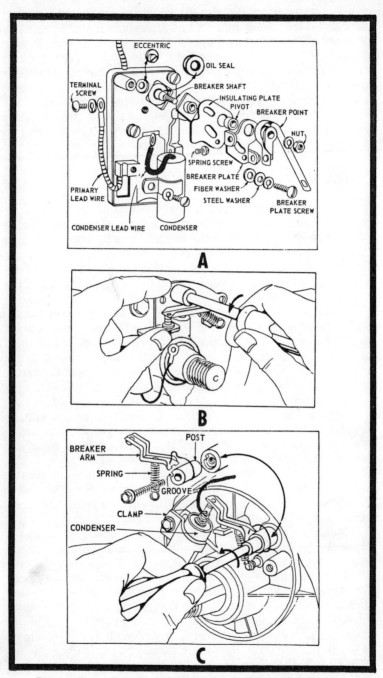

Fig. 2-6. Three Briggs and Stratton Breaker Point Assemblies.

BREAKER POINTS

Other than spark plugs, the breaker points are the most troublesome components in the conventional ignition system. Some sets have an operating life of less than 50 hours. Points are no more than a high speed switch which open and close, interrupting the current flow through the coil

Fig. 2-6 shows three styles of points used on Briggs and Stratton engines. Another variation, not illustrated, is the "saddle points" used on some of the older Clinton engines. These points had a back and forth motion, rather than opening in an arc as in more conventional designs. All point sets have two tungsten or platinum contacts, a moveable arm which rides on a cam, a return spring, and some provision for adjustment.

Troubleshooting the Points

The major cause of point failure is oxidation of the contacts. Oxidation builds up because of arcing in normal operation, and it also builds up during long periods of storage. It is not unknown for brand new engines to refuse to start because of dirty points. Ideally, the points should be replaced, not dressed, because the file leaves tiny irregularities in the contact surfaces which encourages further deterioration. Nevertheless, many servicemen, especially those who are working under a tight budget, file points. For best results, the point assembly should be removed from the engine and held lightly in a vise. The idea is to dress the contacts so that they meet over the widest possible area. Do not use a stone for this operation.

Repair, and Setting Point Gap

Before installing new points, inspect the case for excessive oil. Assuming correct maintenance, oil here means that the crankshaft seal has failed, and should be replaced. The cam should be dabbed with a small amount of high temperature grease and a tiny amount of oil should be put on the point pivot. Try not to touch contacts with your fingers; the acid from your skin will adversely affect (etch) the contacts. Position the point spring carefully, without any unnecessary bending. On most designs, this spring is "hot," that is, it carries primary current. If it is allowed to touch metal, the ignition system will be shorted. All wiring should be tucked out of the way of rotating parts. Align the points by bending the moveable arm so that the contacts meet over their entire surface. Small bending bars are available for this job, but a pair of long-nosed pliers does as well.

Point gap varies from engine to engine. Without the proper data, .020" is correct for nearly all small American engines, and should work in any. To adjust the gap, the crankshaft is rotated until the points are cammed open to their widest position. With the mounting screw slightly loose, move the points into adjustment. Some designs have an eccentric screw while others have a slot for a screwdriver. Either a flat feeler gauge or a wire-type spark plug gauge may be used. An advantage of the latter is that it does not have to be held perpendicular to give an accurate reading. Tighten the mounting screw and recheck the gap. The final operation is to burnish the contacts with a piece of heavy paper (NOT EMERY, JUST PAPER) to insure that they are absolutely clean.

CONDENSERS

Going back to our "aluminum pipe filled with ping-pong balls" analogy for a moment, the condenser (or more properly, the capacitor) can be thought of as a rubber diaphragm in the pipe. The ping-pong balls move against the diaphragm, stretch it, but cannot get past it. As soon as the pressure on the balls is released, the diaphragm springs back, pushing the balls out of the tube. In an ignition system, the condenser functions as a temporary storage space. Primary voltage can reach 300 volts, even in 6-volt battery systems, and could easily arc across and burn the contacts. The condenser is wired across the points to give the arc an alternate place to go. It is easier to charge the condenser than to jump the air gap. Once the condenser is charged, it discharges itself back into the coil. The direction of discharge is the reverse of the direction of the primary and "kills" the magnetic field. The faster the primary field collapses, the greater the induced charge on the secondary, and the better the spark. Thus, the condenser has two functions: to protect the points and to increase the efficiency of the coil.

Condenser Construction

The condenser consists of strips of foil sandwiched between insulating paper. A typical condenser is shown in Fig. 2-7. Half of the coil area is grounded to the case, and the other half is connected via the lead to the hot side of the points.

Electrons, attracted to the ground, crowd together on the hot side, like children at a candy store window. But they are prevented from going to ground by the insulated paper (called the dielectric). Eventually, these electrons will find their way to ground by passing through the primary windings of the coil.

Condenser Faults

Like all electrical components, condensers fail in either of two ways. They can become open or shorted. An open means that the path of current flow is blocked by a high resistance. Usually the high resistance is a faulty ground. Be sure that the condenser is mounted securely to a clean, bright surface. Rust, grease, or paint will cause resistance and poor performance. Opens can also occur at the connection between the terminal and the foil. This kind of fault cannot be repaired. Grounded condensers are the result of a moisture-soaked dielectric or of a puncture.

Troubleshooting the Condenser

Condensers can also leak, that is, be intermittently shorted. If you even suspect it, replace it.

Signs of condenser failure are burnt points (any color other than slate gray should be considered evidence of burning), pitted points, and weak spark. A weak condenser will allow metal transfer from one point contact to the other. The spark should be heavy and sky blue; wiry, white, or reddish sparks usually mean condenser failure. Most mechanics replace the condenser whenever the points are changed. With the current Briggs design, there is no option since the stationary contact is integral with the condenser. But equipment is available to test condensers. Two factors must be checked: the capacitance and the resistance of the dielectric. Most small engine condensers have a capacity of .20 to .22 microfarads and have a dielectric resistance of 3 to 5 million ohms. The condenser should be replaced if the capacity varies more than 15 percent from the manufacturer's recommendation, or if the strength of the dielectric drops to 1 million ohms or less.

It is always easier to replace the condenser with a factory part, but there will be occasions when this is not possible. In an emergency, any name brand condenser having equivalent electrical characteristics may be used (Sprague, Mallory, etc). For battery ignition engines, the replacement should be able to withstand at least 500 working volts, and on magneto systems you should specify 750 or 1000 volts.

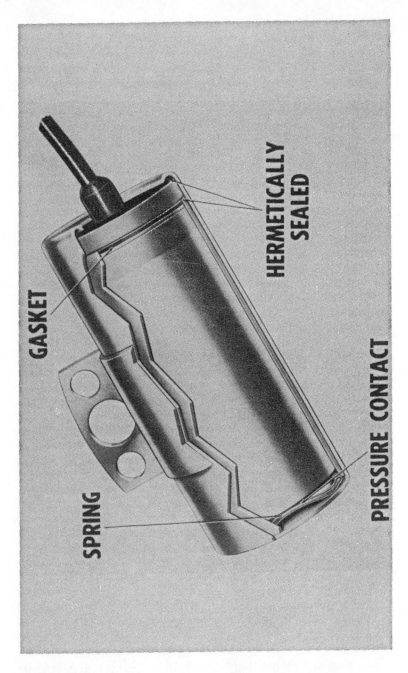

Fig. 2-7. Internal View of Ignition Condenser. (Photo courtesy Delco-Remy.)

DISTRIBUTORS

Engines with more than two cylinders are normally equipped with a distributor. Fig. 2-8 is an exploded view of a typical unit.

The primary function of the distributor is to apportion current from the coil to the individual spark plug leads. Almost universally, the points operate off the same shaft and can be moved, either manually or automatically, relative to the shaft in order to change the timing. Problems can occur in any of these areas. The shaft must be in proper relationship to the engine (which will be discussed later), and should have no perceptible side play in its bearings. The rotor is secured to the upper end of the shaft by a spring clip or a moulded plastic tab. Check for play at this point, for if the rotor is free to turn independently of the shaft, it will be impossible to keep the engine in time. Some rotors have a spring which serves as a conductor of secondary current. It must be clean and in contact with the center terminal. Others use a graphite brush for the same purpose. This brush is subject to wear and shock damage.

Test and Inspection

Inspect the distributor cap carefully. The terminals must be free of corrosion and should be electrically isolated from each other. In other words, resistance between all terminals should be infinite as measured with an ohmmeter. Caps sometimes show a carbon track between terminals where high voltage current has burned a short circuit. Usually, the track will be visible on the inside surface of the cap, but there are occasions when the path of the current is inside of the bakelite. If there has been a short, the cap must be replaced.

Timing

The ignition should be timed so that the full force of the explosion occurs just after the piston has passed top dead center. Since it takes an appreciable time for the fuel-air mixture to ignite, all engines (except "smog motors" with pollution control devices) are timed to fire a few degrees before TDC. Static or fixed advance works fine at any given speed. But an engine which is statically timed for best efficiency at 1,000 rpm will be retarded at 7,000. Regardless of piston speed, the rate of flame propagation remains constant. Some kind of automatic advance mechanism is needed.

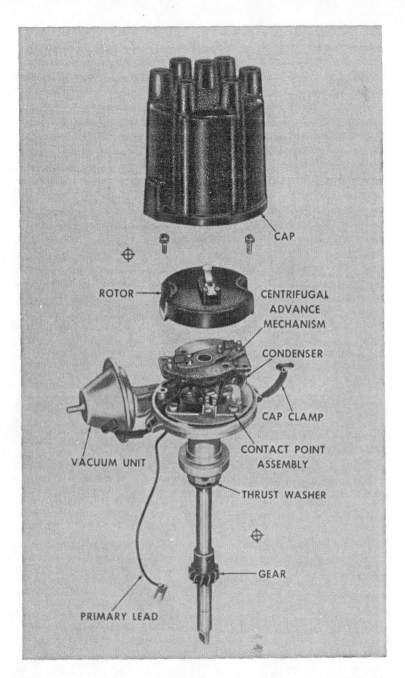

Fig. 2-8. Delco-Remy Distributor Assembly. This distributor, designed for 6 cylinder engines, has all the features found on smaller engines.

Advance Mechanisms

The distributor shown in Fig. 2-8 employs two separate advance mechanisms. One consists of a diaphragm which is vented to the atmosphere on one side and connected to the intake manifold on the other. It responds to engine vacuum. These units can be tested by sucking on the vacuum line. The actuating arm should move and hold its position as long as negative pressure is applied. The second type of advance is centrifugal and depends upon spring-loaded weights which move in proportion to engine rpm. Periodic cleaning and inspection are required for best operation.

Finding the Timing Marks

If the distributor has been removed from the engine and no reference marks have been made as to the position of the rotor and distributor body relative to the crankshaft, the engine will have to be timed. Number one cylinder—counting from the front of the engine—is the key to the whole process. The crank is turned until both valves are closed on No. 1. This means that the piston is on the compression stroke. A screwdriver or a piece of wire is inserted through the spark plug boss and the shaft is rotated until the piston is near top dead center. The exact distance will vary from engine to engine, but $\frac{1}{8}$" is adequate. Next, insert the distributor, making certain that its drive pinion has meshed with the camshaft. Now turn the distributor body until the points just open. You can use an ohmmeter or a piece of tissue paper to get an accurate reading. Install the rotor. It will be aimed at one terminal on the cap; this will be No. 1 cylinder.

The remaining leads are connected as the firing order dictates. On four cylinder engines, the order may be 1243, 1342, or 1432. Remember that you must also take into account the direction of rotation of the rotor.

At this point, we have discussed the major ignition components. But these components are only so much hardware if they are not connected together as a team. Now let's look into complete ignition systems.

BATTERY IGNITION SYSTEM

The oldest system, and one that is used on millions of automobiles, battery ignition is probably the easiest to diagnose and repair. Look at Fig. 2-9. Primary current, supplied by a battery, flows through the coil and to the points.

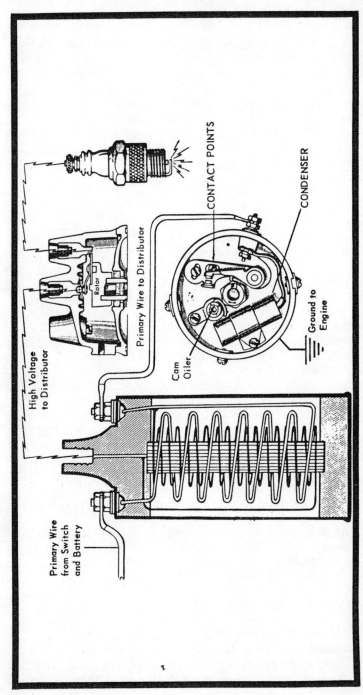

Fig. 2-9. Typical Battery Ignition System.

As the cam rotates- this particular drawing is for a six cylinder engine—the points open, interrupting the flow of primary current and producing high voltage secondary current as the primary field collapses. The high voltage is distributed to the spark plugs by the rotor and cap assembly. Single cylinder battery ignition systems are similar, except that the distributor is unnecessary and is omitted.

Troubleshooting

First ascertain that the battery is fully charged. (The chapter on electrical systems goes into some detail about the procedure used to check batteries.) Be certain that the battery terminals are clean and bright and that the battery has a good ground; there is very little voltage on the primary side—six or twelve volts can be blocked by paint or rust. Next, trace the circuit to the ignition switch and see that battery voltage is getting to one terminal. With the switch closed, both terminals should be hot. Move to the coil to determine if the primary lead is hot at the terminal. Make the same check at the distributor terminal. Examine the points carefully. If they are pitted or blue or black, replace them. Most of the time, the points are at fault in any conventional ignition. Set them according to the manufacturer's specs and crank the engine. If the current is getting through the contacts, they will spark as they open and close.

Further Tests

Suppose that all of the above checks have been made, and there is still weak or no high tension current. In such cases, replace the condenser and recheck the points. On distributor equipped engines, disconnect the center wire at the cap. Hold it ⅜" from a good ground at the block. If there is spark at this terminal, but none coming out of the spark plug terminals, the distributor or the spark plug leads must be at fault. Two things can go wrong with the high tension side of the distributor: either the rotor may be physically damaged, or else the cap may have become cracked and-or shorted.

If there is no spark at the coil, then by elimination, the coil itself or the high tension lead wire from the coil must be at fault. Metallic wires rarely give trouble, but the radio suppression types should be tested, preferably by substitution.

Experienced mechanics do not normally make a step by step analysis as just outlined. They know a few "short cuts" and have a good idea of what to expect. Their procedure is to

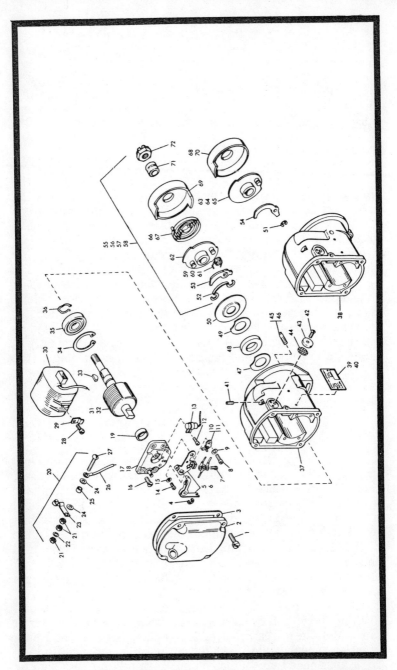

Fig. 2-10. Typical External Magneto. Parts numbered 51 through 72 include 2 variations of Impulse Coupling. (Courtesy Fairbanks-Morse.)

check the low tension side by seeing if the points spark and then to test the high tension at the coil, and work from there.

Routine maintenance consists of keeping all the contacts clean (including those on the distributor cap), periodically replacing the points and condenser, and timing the spark. Nonmetallic cables should be checked for high resistance.

MAGNETO IGNITION SYSTEMS

Most small engines employ magnetos to "light" the spark plugs. The main advantage of the magneto is that it is a self-contained unit, requiring no battery. Since primary current is supplied by permanent magnets which saturate the coil with moving magnetic lines of force, the engine can be stored for months without any maintenance or bother about keeping the battery charged. Neither is there the complexity of a generator or alternator to charge the battery while the engine is operating. Magnetos also have the virtue of producing a hotter spark in nearly a direct proportion to engine rpm. Battery systems tend to produce less voltage at high speed, when it is needed the most. But all engineering is a compromise. The advantages of magnetos are counterbalanced by a number of factors.

Inherent Problems of Magnetos

Since the magneto depends upon the speed of rotation to develop spark, high tension voltage will be relatively weak at low rpm. This can be critical for starting. Some magnetos, as used on industrial engines and aircraft, have an impulse coupling (Fig. 2-10) which kicks the armature through a half turn by spring tension. BMW has gone so far as to use a battery for starting and a magneto that cuts in for running. Small, single-cylinder engines usually do not have special provision for starting. It is assumed that the operator can spin the engine fast enough.

Another problem with most magnetos is high primary voltage. This voltage tends to arc the points and does put a stress on the coil primary insulation. Heat is another problem: a battery coil can be placed anywhere on the engine, the only limitation being the length of the leads. Magneto coils are mounted on the engine block, and often under the flywheel.

And from a mechanic's point of view, magnetos are difficult to service. Several factors are responsible. Unless the engine is running, there is no primary current. In our discussion of test procedures on battery ignitions, we used

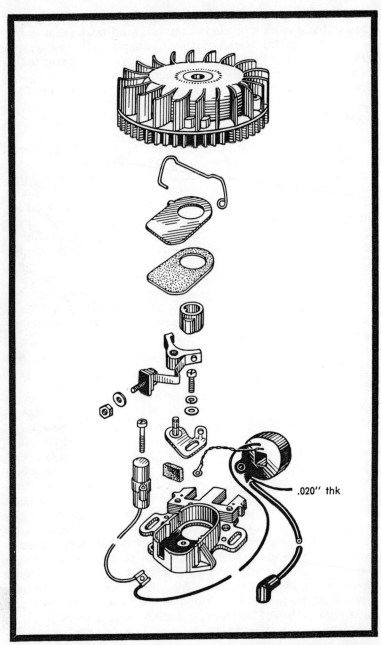

Fig. 2-11. Flywheel Magneto for 1-cylinder Engine, without auxiliary lighting coils. Note that tab on the point chamber case is exactly .020" thk, and can be used to set point gap. (Photo courtesy R.E. Phelon Co.)

primary current as a kind of base line which showed the mechanic that at least half of the system was functioning. Another problem is that magnetos are designed within definite space limitations. The parts are closely packed together, and one must be careful with the routing of wires so that they do not touch moving parts such as the crank or the flywheel. Parts nonavailability also is a problem. Battery coils and condensers of different manufacture are fairly interchangeable; magneto components are usually designed for a particular engine series and do not interchange. This may even be true of mechanical components such as flywheels. I remember one engine which arrived in the shop as a basket case, but all the parts were there, and we assembled it. No spark. We successively replaced the points, condenser, the coil, tested the magnets, checked all the connections, and still no spark. Finally, someone noticed that the flywheel was not the same part number as the one called for on that particular engine. It fit perfectly, but the magnets on the rim were in the wrong position relative to the coil and the crankshaft key.

Servicing Magnetos

Before we get into discussions of detailed servicing, I suggest you study Figs. 2-10, 2-11, 2-12, and 2-13 thoroughly, to obtain an overall view of parts and locations on typical magnetos.

Most small magnetos are the flywheel type. That is, the flywheel carries the magnets and coil, while the condenser and points are located under the wheel. The first step in servicing these magnetos is to remove the wheel. On some engines, the shrouding must come off first, and on small motorcycles, it is usually necessary to remove the primary chain cover. The flywheel is fixed to the tapered end of the crankshaft by means of a key and a nut. Remove the nut—normally it will have a right-hand thread, but certain Briggs engines without a rewind starter used a left-hand thread. After the nut is removed, the flywheel will still be in position, held by the key and the tapered fit of the crankshaft. At this point, you should use a flywheel puller. Each engine manufacturer supplies pullers for his own engines. These tools enable the mechanic to remove the wheel without warping it and without hammering on it. Vibration can damage the magnets.

Pulling the Flywheel

But even if the correct puller is available, most mechanics do not use it. They work under time pressure, and will reach

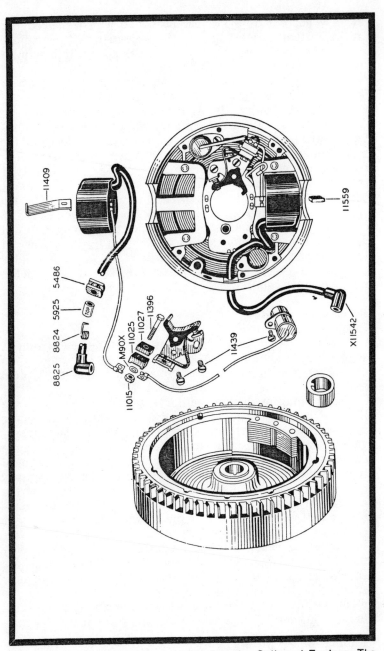

Fig. 2-12. Flywheel Magneto for Twin-Cylinder Outboard Engines. The duplication of parts makes it easy to perform substitution checks. (Courtesy Wico)

for a "knocker." This is a heavy bar which is threaded to match the crankshaft; for most American engines, you need ½" and 5-16" sizes. The Briggs rewind starter (see the chapter on power transmission for illustrations) uses a crank which is unthreaded and chamfered at the end. Knockers for these engines are made by boring a half inch hole in the steel bar to a depth of two inches. The finish cut is made with a squared-off bit, so that the end of the hole will be flat; a round-ended hole would jam on the chamfer of the crank.

The knocker is run down on the threads almost to the flywheel and then struck with a heavy hammer. A screwdriver is sometimes placed behind the rim of the wheel to force it outward. A word of caution: using a knocker may be convenient, but it can cause trouble. Do not hit the shaft with a glancing blow. Most small engine shafts are cast iron and will snap. If the wheel is particularly stubborn, try penetrating oil. Stay out of the way if you're not doing the work; mechanics have been known to drive the shaft right out of the opposite side of the engine block.

On Japanese and European engines, it is common practice to use the flywheel nut as the knocker. The nut is run back to the end of the threads and struck with a lead hammer. This practice is not recommended. Any shop which accepts a job should charge enough to buy whatever special tools are required to do the work.

Detailed Inspection and Testing

Inspect the keyway for wear and for cracks in the hub and on the shaft. Some aluminum wheels have a steel hub which is riveted into place. See that these rivets are tight. The smallest misalignment of the wheel to the crank can disable the magneto. The magnets and the points are timed at the factory to extremely critical tolerances.

In many rotary lawnmowers, the blade is an auxiliary flywheel. Unless it is securely mounted to the shaft, the engine will backfire and may refuse to start.

The magnets themselves rarely give problems. Years ago, before Alnico and other exotic alloys were developed, it was common practice to "re-charge" the magnets. Today, this is rarely necessary, and few shops have the equipment to do it. You can estimate the strength of the magnets by comparing them to new ones. If they are weak, the best solution is to replace the flywheel. However, this is not always possible on old or foreign engines. Bomar Magneto Service, 2546 Cherry Avenue, Long Beach, California, 90806, specializes in this work and carries on an extensive mail order business.

The ends of the armature should be close to, but not in physical contact with the magnets. The closer this E-gap is, the hotter the primary voltage. Rubbing will cause overheating and power loss. The gap should be between .006" and .010" on the larger engines. Usually the armature mounting screws have slotted holes which allow the assembly some fore and aft movement. A strip of cardboard from a postal card can be used to space the armature.

Oil in the magneto housing means that the crankshaft seal should be replaced. The points will not "live" in an oily environment.

Troubleshooting flywheel magnetos is a matter of elimination for most mechanics. First replace the points and the condenser. Install the flywheel and spin the engine with the plug out. There should be a heavy, blue spark which will have enough voltage to jump $3/8$" or so to ground (the block). If the spark is weak, the next logical choice is the coil. First see that its ground wire is securely connected and that the high tension terminal is clean. Without special instruments, it is impossible to measure the primary current on most magneto designs. Sometimes an engine will run for a few minutes and then stop. Usually, the problem is with the coil, although condensers have been known to short out with heat.

Stop switches and "kill buttons" should also be checked. The latter are found on motorcycle handlebars and are usually carbon contacts which can disintegrate, and short out the ignition.

A variation of the flywheel magneto is the rotor type. Assembled, it appears identical to the flywheel type, but has one important difference. The magnets are fixed to a rotor and not the wheel. These designs are found on outboards and the larger industrial engines. Often the rotor is not keyed to the crank and can be rotated once the pinch bolt is loosened. Adjustable rotors always have a timing mark which aligns with one on the stator plate. These marks should be in line at the moment when the points open. Otherwise, maintenance and troubleshooting is the same as for conventional flywheel magnetos.

SPECIAL IGNITION SYSTEMS

Below are descriptions of newer types of ignition systems coming to the forefront, with emphasis on special testing techniques.

Energy Transfer Systems

Pioneered by Wico-Pacy Lucas, these systems are found on most large British and some Japanese motorcycles. The design is similar to, although much more sophisticated, than the low-tension magnetos of grandfather's day. Basically, it is a rotor type magneto with two coils. One is mounted under the flywheel and the second coil is outside of the engine. The first coil produces low tension energy that is amplified by the second. The big advantage is that the second coil can be as large as necessary, since it is remote from the flywheel, and that it can be mounted wherever convenient.

Test procedure is to substitute a 6-volt battery (two batteries for the twin cylinder models) for the generating coil. Disconnect the primary lead—coming from the flywheel—of the external coil and connect one side of the battery to the coil and the other to the frame. Observe correct polarity. Working quickly, because the external coil is not designed for battery operation and will overheat, crank the engine. If there is spark at the plug, the points, condenser, and the external coil are OK. The fault must lie with the generating coil or in the rotor. Individual components are checked out as explained previously.

External Magnetos

Some of the better industrial engines and a few of the older motorcycles feature external magnetos. In the United States, the two leading manufacturers are Fairbanks-Morse and American Bosch. Maintenance is the same as outlined previously, except that the magneto drive must be timed to the engine or assembly. Different models use different methods, but basically the troubleshooting drill is the same as for distributor timing. Many external magnetos have an impulse coupling which automatically retards and intensifies the spark during starting. The coupling consists of a pawl, a stop pin, and a coil spring as in Fig. 2-10. At low rpm, the pawl engages the stop which coils the spring for part of the revolution. The pawl is released and the spring suddenly unwinds, snapping the magneto shaft into firing position. Once the engine starts, centrifugal force retracts the pawls away from the stop pins. Maintenance problems are rare, but the pawls and stop pins can wear and the springs have been known to break from fatigue.

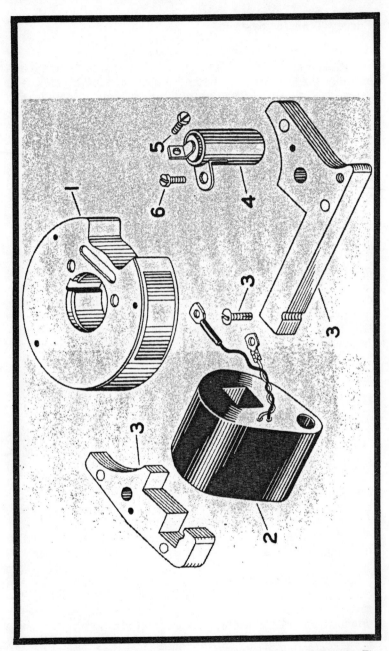

Fig. 2-13. Repco Rotor-Type Magneto used on Kohler K161 Engine. The rotor is keyed to the shaft; but in other designs, the rotor position can be moved to change the timing. (Courtesy R.E. Phelon Co.)

Capacitive Discharge Systems

Two years ago, Champion tested the major brands of motorcycles for spark output. They discovered that the average ignition system developed between 12,000 and 15,000 volts at the spark plug terminal. For best performance, this figure should have been in the neighborhood of 20,000 volts. Less than this means that the spark plugs will short out when the electrodes become fouled with carbon deposits. The Capacitive Discharge Ignition, or CDI as it is familiarly called, is the best solution to the problem to date. A good CDI can produce 35,000 to 50,000 volts and do it quickly with a relatively small amperage drain. Some CDIs are so potent that they will fire plugs which have been dipped in oil.

CDI is only remotely similar to battery or magneto systems. In the first place, there are no points in these new ignitions; all switching is done electronically. The elimination of the points is, in itself, a big plus. Points are liable to burning—especially when switching high amperages—and are prone to the ills that affect any moving part. Friction, wear on the rubbing block, point bounce at high speed, can all be staved off by good engineering, but not completely overcome.

In the early 1960s, automobile designers experimented with a hybrid system. The points in a conventional battery system were supplemented with a transistor. This solid-state device, no bigger than a fingernail, is a high speed switch or gate. A small current impressed on one terminal can "open" the gate to allow large currents to open in a second circuit. Another way of saying this is that a transistor is a relay with no moving parts. The points now could carry very little current—just enough to activate the transistor—and, consequently point life increased dramatically. A further development was to eliminate the points entirely, by replacing them by a coil and a rotating magnet. As the magnet sweeps past the coil, a small current is generated which is fed to the transistor gate. Some of these hybrid systems are still on the market.

But here is the problem with them and the reason why they were never used on small engines: to get high voltage, the primary windings of the coil must carry heavy amperage. This current means larger, heavier coils and a greater load on the electrical system. Small engine designers were not willing to pay this penalty merely to do away with the points. So matters stood until the advent of the CDI in the mid-sixties.

A capacitor is nothing more than a large condenser of the type found wired across the points in conventional systems.

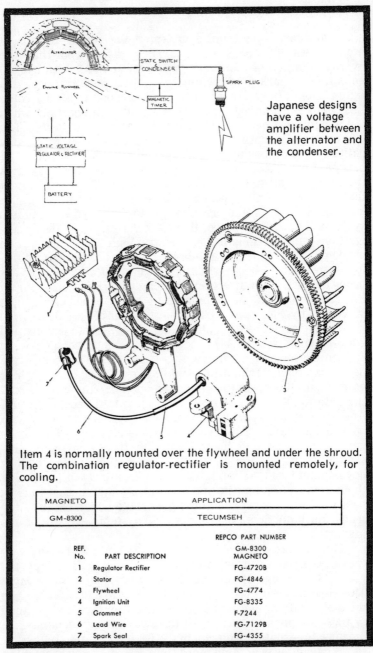

Japanese designs have a voltage amplifier between the alternator and the condenser.

Item 4 is normally mounted over the flywheel and under the shroud. The combination regulator-rectifier is mounted remotely, for cooling.

MAGNETO	APPLICATION
GM-8300	TECUMSEH

REF. No.	PART DESCRIPTION	REPCO PART NUMBER GM-8300 MAGNETO
1	Regulator Rectifier	FG-4720B
2	Stator	FG-4846
3	Flywheel	FG-4774
4	Ignition Unit	FG-8335
5	Grommet	F-7244
6	Lead Wire	FG-7129B
7	Spark Seal	FG-4355

Fig. 2-14. Repco-Spark Capacitor Discharge Ignition (CDI) System. (Courtesy R.E. Phelon Co.)

Earlier in this chapter, the analogy was made between a capacitor, or condenser, and an elastic diaphragm placed in a pipe. As the ping-pong balls are packed into the pipe, the diaphragm is stretched. When pressure or voltage is taken off the balls, the diaphragm springs back into position, discharging the balls back out the pipe. This analogy assumes that the flow of current (the direction of the balls) is constant. They move from the end of the pipe to the diaphragm. This is direct current. Suppose that the direction alternated in the conductor—right to left, then left to right. The diaphragm or capacitor would be useless. Before the diaphragm was stretched or charged, the direction of flow would change. It might as well not be in the circuit.

All of this is an attempt to explain why capacitor discharge systems must operate on direct current. Magnetos are AC devices. The current from them alternates back and forth in the conductor. Therefore, CDI must have some sort of rectifier which converts AC to DC in order to charge the capacitor. These rectifiers are solid-state devices which allow current to go one direction and block it in the other. The chapter on electrical systems goes into more detail on the matter.

Now let's put these components together. Alternating current is generated in the field coils in the magneto and flows through the rectifier. The current, now moving in one direction, charges a large capacitor. The transistor is triggered by a coil, excited by a magnet on the crankshaft. That is, as the shaft turns, a magnet excites this coil, and the transistor opens, allowing the current collected in the capacitor to discharge through the primary side of the coil.

The accumulated energy produces a whopping voltage on the secondary side of the coil. And the voltage build-up is fast. Compare the rise time for conventional systems which averages from 8 to 125 microseconds (a microsecond is one millionth of a second). CDI systems do it in 2 microseconds. This means that a badly fouled plug will still fire, since the current is pouring in so quickly that it does not have time to leak away through the carbon deposits. Conventional systems also have the unfortunate tendency to erode the gap between the spark plug terminals. This is because the high tension current is, at the moment of spark, moving in one direction. Metal migrates from one electrode to the other in the direction of current flow. The CDI current is alternating and there is no metal migration. Consequently, plug life is increased.

Before we leave the subject, there is one refinement which should be mentioned. The capacitor works best at 200-400 volts. Ossa, Bultaco, Repco, and Kokusan units use a special coil in the magneto to generate this voltage. Other systems are supplied with 6 to 12-volt current (from the magneto or from the battery), which is then boosted to the required level. In the two and three cylinder Kawasakis, this boost is created by a transistorized oscillator, which is similar to the power supply on a car radio.

CDIs are not repairable in any real sense. The coils themselves are replaceable, but the capacitors, switching transistors, rectifiers, oscillators, and the rest are encapsulated in silicon or epoxy. These "black boxes" give good moisture and vibration protection, but must be replaced as a complete unit. Theoretically, it is possible for a capable electronics technician to "jimmy" one of these boxes and replace the individual components which are available from radio supply houses. However, this is not done in commercial engine repair shops.

Troubleshooting is normally done by substitution or by factory test equipment. If routine maintenance such as removing rust from the magnets and cleaning all contacts (the transistor signal circuit is especially vulnerable) does not help, it is better to farm these jobs out to a small engine parts distributor. Substitution at $80 or more per component can get mighty expensive!

Chapter 3

Fuel Systems

The fuel system is second to the ignition in the frequency of service problems. Most of these problems involve contaminated fuel which clogs the carburetor jets. Gasoline has an open shelf life of less than six months, and once allowed to go "stale," it becomes corrosive. It causes synthetic rubber gaskets and fuel lines to swell and disintegrate. Unless the climate is arid, stored gasoline will pick up water from condensation. Water causes the tank to rust and will attack the zinc alloy carburetor castings. Sometimes the carburetor will be plugged solid with crystals of zinc oxide.

Another problem with the fuel system is leakage. Leaks may develop at the tank (especially around the mounting fixtures), the fittings on or in the line, or near the ends of the fuel line itself. In a few cases, leaks are caused by physical damage, but most of them develop slowly over time as the result of vibration and fatigue. Internal carburetor damage such as a stuck or punctured float can also cause fuel leakage. Air leaks become an immediate problem when they occur on the engine side of the carburetor. Most commonly, these leaks are caused by faulty gaskets at the carb mounting flange or at the intake pipe-cylinder joint. Two-cycles may leak at the crankcase seals and four-cycles, in time, will develop leaks around the valve guides. Another leak area is around the throttle shaft or, on Amal-type carbs, around the throttle slide.

Air leaks usually show up at low rpm. The engine will be difficult to start and may refuse to idle. An examination of the spark plug might show a lean condition (unless someone has over-jetted the carburetor in an attempt to cure the leak by treating the symptoms). In severe lean conditions, there will be signs of detonation and pre-ignition. Mechanics are generally content with a visual inspection and a routine replacement of gaskets. In "hard cases" where the engine appears to be leaking, but the location of the leak cannot be found, it is helpful to use a stethescope. These instruments can be found on the military surplus market or can be purchased from mail order houses. Another method is to squirt small quantities of gasoline on the suspected joint as the engine is running. Be extremely careful around hot parts such as the exhaust manifold. Theoretically, if the engine is leaking air, it should speed up as gasoline is squirted near the leak and is pulled in to the engine.

Two-cycle seal leaks are another matter entirely. On a single-cylinder engine, one or both seals will be inaccessible for direct test. On the multi-cylinder engines, the seals between each cylinder are only visible when the engine is stripped. Most shops replace seals by deduction: if all other possible sources of trouble have been eliminated, then the seals must be at fault. It is possible to remove the cylinders and carburetor, seal the engine off from the atmosphere, and to run a pressure test. The chapter on engine service, goes into more detail on this procedure.

Insufficient, or the converse, excessive fuel delivery can also be caused by misadjusted carburetor controls. If the main jet is screwed shut, the engine will starve. Suspect that any jet—main, intermediate, or idle—which is readily accessible has been tampered with.

In sections that follow, various components of the fuel system are described generally, with step-by-step, logical procedures to use when isolating problem areas.

GENERAL TROUBLESHOOTING PROCEDURE

The prime indication of trouble in the fuel system is a dry cylinder after repeated cranking. But merely because the plug is dry, do not assume that the fuel system is at fault. For this symptom to be meaningful, a number of criteria must be meant:

1. Is the engine cold? A hot chamber will evaporate the fuel charge.

2. Does the engine have sufficient compression to pump into the chamber? Most small engines require at least 60 pounds per square inch, as measured on a compression gauge, to start.

3. On two-cycles, is there sufficient crankcase pressure to pump fuel through the transfer port? Crankcase pressure is difficult to measure (it can be felt as some engines are cranked through with the spark plug removed). And pressure can escape through bad seals or through a reed valve which is not seating.

4. Is there fuel in the tank and are the fuel petcocks open? This seems elementary, but it happens frequently.

The second indication of fuel system trouble is too much fuel in the cylinder. Too much gasoline will "flood" the plug and short it out. Flooding is the result of a leaking float valve, a punctured float, or a choke butterfly which is not opening. On two-cycles, persistent flooding is possible if the crankcase is flooded. The latter is caused by tilting the engine, or by a fuel petcock which was left open.

SYSTEM TROUBLES

If fuel does not flow from the tank, suspect the petcock screen. Fig. 3-1 shows one of these valves as typically found on motorcycles.

If motorcycle valves leak, it is best to replace the unit, rather than to attempt a repair. Industrial engines have a simple screw type needle valve, often in combination with a glass fuel bowl. These valves may be ground with valve lapping compound to reseat them and can be repacked with string packing. Wisconsin dealers are a good source of this packing. The fuel bowl gasket is usually made of cork and will shrink when dry. Do not remove the bowl more than necessary for cleaning.

Fuel Lines

Most fuel lines are made of synthetic rubber or of neoprene. An advantage of neoprene is that you can see the fuel flow. Aluminum or copper tubing is sometimes used, although it is prone to fatigue cracks at the fittings. If a long (eight inches or more) section of metallic tubing is used, it should be coiled three or four times between fittings. The coils act as a spring and soak up some of the vibration.

When replacing flared lines, most mechanics prefer to use ferrule type fittings. These fittings make their own flare as they are tightened. Automobile supply houses carry a wide range of sizes. If a flaring tool is used, extend the tube approximately 1-16" above the holder. This is to insure enough material to make a leakproof flare. It is a good idea to wrap all jointed connections with teflon tape.

Fuel Pumps

Small industrial engines are often equipped with a mechanical pump similar to the one shown in Fig. 3-2. The pump arm rides on the camshaft and is actuated on every second revolution of the engine. Gasoline to the pump and no

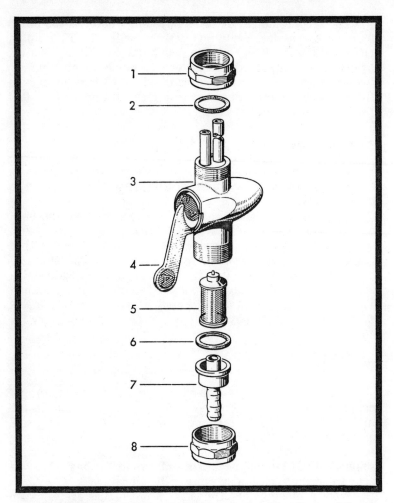

Fig. 3-1. Fuel Petcock used on BMW Motorcycles. The shorter of the two inlet pipes feeds the reserve setting. (Courtesy Butler and Smith, distributors for BMW.)

discharge means, of course, that the pump has failed. Check the valves first, since a small particle of rust can jam one open. On some pumps, it is possible to install the valves upside down. During the actuating stroke, the inlet must close and the outlet must open. The next most likely part to fail is the diaphragm. Sometimes you can see that it has cracked, and other times it will have stretched. When in doubt, replace it. After long usage, the arm and the pivot pin on which the arm swings can wear. In some cases, it is possible to "resurrect" the pump by welding a tab on the end of the arm, but most

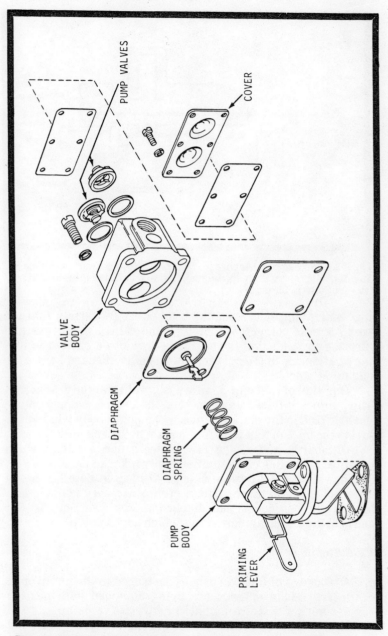

Fig. 3-2. Mechanical Fuel Pump used on Kohler K662 Engine. When installing this particular pump, hold the priming lever down until the mounting screws are fully drawn up. Installation of any mechanical pump is easier if the actuating lever rides on the heel of the cam. (Courtesy Kohler of Kohler.)

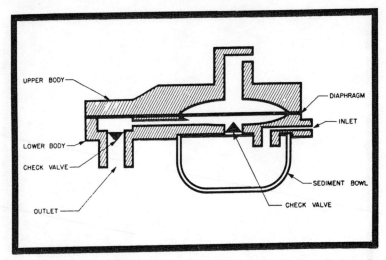

Fig. 3-3. Schematic of Diaphragm Type Fuel Pump used on Outboards. The diaphragm responds to fluctuations of crankcase pressure. (Courtesy McCulloch Corporation.)

mechanics simply replace the whole unit. Be certain that the pump arm is correctly located on the cam lobe. Briggs and Stratton suggests that the engine be torn down enough so that the relationship of the parts can be visually checked before the engine is started.

One sign of fuel pump failure is gasoline in the crankcase. The pump will flow enough to keep the engine running, but a pin hole leak in the diaphragm or in the pump body will bypass fuel to the oil. Fortunately, this is rare, but check for it.

Outboards and riding lawnmowers use a pump which operates by means of fluctuations in crankcase pressure. Fig. 3-3 shows a typical design which will vary in detail between different manufacturers. These pumps are very reliable and only require an occasional diaphragm replacement. As you can see, the diaphragm has flaps which act as valves.

Fuel Stoppage

A blockage of fuel can usually be traced to the pump or to the filters. Many engines are being equipped with pleated paper filters which are mounted between the pump and the carburetor. These devices can be checked by gently blowing through them. Cleaning is not practical, and the whole unit should be replaced. Fuel screens which are placed at the tank outlet, the fuel pump inlet, and (rarely) at the carburetor, can be visually checked and cleaned. Bronze filters consist of tiny spheres of bronze molded into shape; don't clean them, replace them. And finally, the fuel supply will be cut off if the tank cap is not vented. The vent may stop up or the wrong cap may have been used. Some tanks will accept radiator caps. The engine will run a few minutes until the fuel level has dropped enough to cause a partial vacuum in the tank. Air in the float chamber, which is at atmospheric pressure, blocks the flow of fuel. The engine will restart as soon as the cap is removed.

CARBURETORS

The job of the carburetor is to mix fuel and air in the proper proportions. If engines ran at a constant rpm, carburetors would be no more complex than perfume atomizers. But modern engines are designed to operate over a wide speed range. Some Hondas, for example, can tick over at 750 rpm at idle, and develop their maximum power at more than 10,000 rpm. With each change of speed, the engine requires a different proportion of fuel and air. At idle, the ratio is one pound of fuel to eleven pounds of air: at part throttle, the ratio changes to one to sixteen; and at full throttle, it drops to one to twelve. In order to start a cold engine, the ratio must be in the neighborhood of one to six. These ratio changes account for the complexity of modern carburetors.

Principles of Carburetion

All carburetors have a restriction in the bore called a venturi. See Fig. 3-4. Incoming air must pass through the venturi on its way to the engine. As the air stream meets the restriction of the venturi, its speed increases and air stream pressure drops. That is, there is a partial vacuum around the venturi. (The same principle is used in the design of airplane wings—the curved upper surface of the wing is a kind of two-dimensional venturi that creates a low pressure area which lifts the plane.) The fuel, on the other hand, is under atmospheric pressure of approximately 14.7 pounds per square inch. It flows through the main or high speed jet to fill the partial vacuum in the venturi. Once in the air stream, the fuel atomizes into a vapor which can be efficiently burned by the engine.

The venturi principle depends upon the speed of the engine. At low speeds, there isn't much vacuum, and at starting, hardly any. So carburetors are normally fitted with some sort of fuel enrichment device which comes into play during starting. Most often it is a butterfly valve in the bore

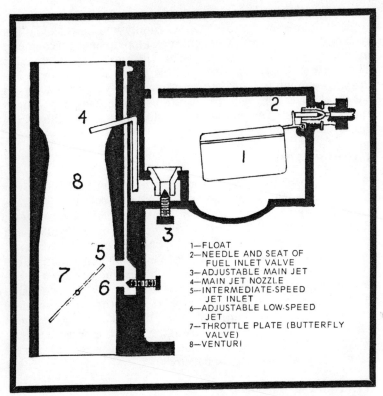

Fig. 3-4. Basic Parts of a Float Type Carburetor. (Drawing by Rene Manuel.)

above ("upwind") of the venturi. When the choke butterfly is closed, the engine is pumping on a blind pipe, and there will be high vacuum in the carburetor and consequently a high rate of fuel flow. Other designs employ a small pump to richen the mixture; in some cases, they have a special jet just for starting.

Because of the limitations of the venturi, most carburetors have an auxiliary jet for low-speed operation (Fig. 3-4). The low-speed, or in British parlance, the pilot jet, will be located near the throttle butterfly and it will operate from the vacuum caused when the valve is partially closed. As the butterfly opens, there is less of an obstruction in the bore, less pressure drop, and the low-speed system cuts out. Many carburetors have an intermediate-speed jet to smooth the transition from low to high speed. It is located just behind the throttle butterfly. As the butterfly opens past the throttle position, the intermediate jet is uncovered and fuel flows.

Thus, most small engine carburetors have three points of entry for fuel: the main jet for high-speed operation, the intermediate-speed jet for the middle range, and the low-speed jet for idle. Except in the simplest carburetors, the main and low-speed jets are adjustable.

There is one other feature of carburetors which we should mention. Many are provided with an airbleed to the main or low-speed jet. The idea is to allow a small amount of air to mix with the fuel prior to entry in the carburetor bore. This airbleed breaks the fuel up into globules which can more easily be vaporized.

Small engines have a great variety of carburetor designs which reflect the range of uses of these engines. A lawnmower engine can get by on a very simple carburetor, since operation over a wide speed range, and fuel economy, are not critical factors. On outboards, fuel consumption is of paramount importance, especially at cruising speed. A motorcycle designer may opt for a wide speed range and singular displacement flexibility, and a chainsaw demands a carburetor that functions at large angles to the horizontal. Another factor is that there are many firms which build carburetors, in addition to the engine manufacturers themselves. Each firm has its own favorite solutions to problems, and each operates under its own more of less unique patents.

Suction Lift Carburetors

These are by far the simplest design and have been around since World War I. Sometimes erroneously called "mixing valves," these carburetors are found on some of the smaller Clinton, Tecumseh, and Briggs engines (Fig. 3-5 shows a Briggs design which is used on millions of lawnmowers.) These carburetors are always fastened directly to the fuel tank. A pickup tube extends downward from the carburetor body into the tank. On the end of the tube there is a fine mesh screen, which serves as a filter, and a ball check valve. There is usually an idle adjustment screw to limit the movement of the throttle butterfly, and an adjustable main jet. Some designs have a low-speed needle, but most have fixed jets, as does the carburetor illustrated in Fig. 3-5. Most suction-lift carburetors are made of zinc-based "pot metal," although a few have been constructed of welded sheet steel.

Because of their inherent simplicity, these devices are quite reliable. The major problem occurs when fuel is left in the tank over a long period, such as during storage over the winter months. Varnish from the gasoline will clog the ball

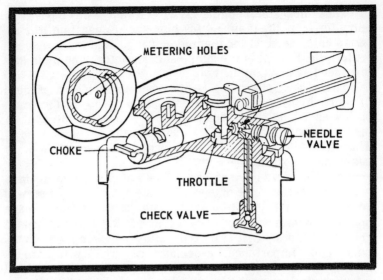

Fig. 3-5. Simple Suction-Lift Carburetor. Other designs (e.g., Clinton) may have an adjustable low-speed jet. (Courtesy Briggs and Stratton.)

check valve on the end of the pickup tube. The repair is either to replace the intake tube with a new one, or to free the valve by inserting a fine wire through the screen and pushing upward. Pickup tubes are generally an interference fit with the carburetor body. The tube is twisted counterclockwise and pulled outward until it comes free. New tubes should be installed (with the aid of a vise) to the depth of the original. On current Briggs designs, the tube should extend below the carburetor mounting flange 2 9-32" to 2 5-16".

Cleaning the Suction-Lift Carburetor

The main jet should be removed—it is visible after the needle is unscrewed—and cleaned with compressed air.

CAUTION: For safety's sake, the compressor should be regulated to supply no more than 30 pounds per square inch. This is enough pressure to clean carburetors and not enough to penetrate the skin and send air bubbles into the bloodstream.

Fixed jets which are located near the throttle plate may be cleared with air or with a soft bristle. Do not use wire on any carburetor jet, as these parts are critical. Inspect the needle for bending at the tip. It must be true for the engine to run efficiently.

Leaks may develop around the junction of the carb body and the tank. After checking the tank surface for cracks, which will usually radiate from the mounting screw bosses, replace the old gasket with a new one. If the parts are distorted from improper tightening, use two gaskets. Leaks are especially dangerous on these engines since most of them are constructed so that the exhaust exits right over the tank.

Adjusting the Suction-Lift Carburetor

Tuning is easy. With the tank at least half-full of fresh gasoline, allow the engine to warm up to its operating temperature. One and a half turns open on the main jet needle will supply enough fuel for the engine to run. Blip the throttle butterfly to full open—the engine should pick up speed smoothly; if it hesitates or stumbles, the main jet is too lean. Open it an eighth of a turn and test again. An overly rich setting is indicated by thick, sooty smoke in the exhaust at full throttle. Single-adjustment carburetors such as the Briggs design may be rich at idle if the high speed mixture is correct. Little can be done about this built-in design limitation. In 99 cases out of 100, the customer will not know the difference. Other makes, such as the Clinton, do have low-speed needles. For these, the main jet is adjusted as indicated above and the low-speed jet is then turned to give the fastest idle. It may be necessary to adjust the idle speed screw on the throttle butterfly. Idle speed should be at least 600 rpm. Some shops like their customers' engines to tick over at 200 rpm or 300 rpm. This sounds good and impresses the customer—however, it is an invitation to a thrown rod on four-cycles. If the throttle is suddenly opened, there may not be enough oil on the crank pin. It is not uncommon for one of these "Cadillacing" engines to be brought back to the shop, the same day it was picked up, with a broken rod. The shop has the choice of replacing the rod and crank free of charge or losing the customer.

Float-Type Carburetors

These types are more sophisticated than the suction-lift types. Figs. 3-6, 3-7, and 3-8 show three styles which are found on most of the small engines of the world. The updraft design is the oldest and has the advantage that it does not increase the height of the engine. On the other hand, starting may be difficult, since the fuel-air mixture has some considerable distance to travel before it enters the chamber. Another disadvantage is that the mixture must make a 90 degree turn

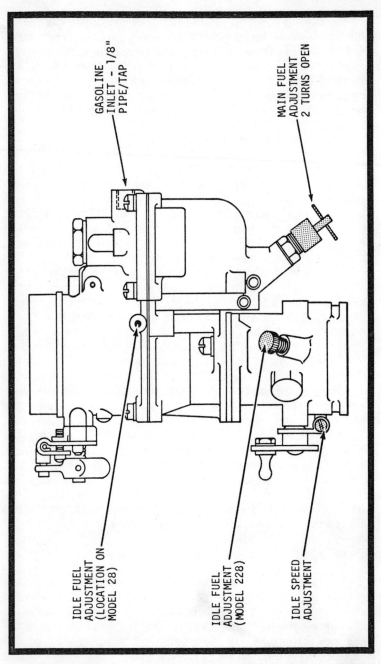

Fig. 3-6. A Downdraft Carburetor, Showing the Adjustment Screws. (Courtesy Kohler of Kohler.)

on its way into the chamber. On low-speed, industrial engines, this is of no consequence; but on high-speed engines, it does put a sharp limit upon the power available.

Down-draft carburetors, such as the one pictured in Fig. 3-6, are quite efficient, but increase the height of the engine. Motorcycles and outboards are almost always equipped with side-draft carburetors such as the Lawnboy type shown in Fig. 3-7. A small disadvantage of both the side- and down-draft types is that when they're flooded, fuel dribbles into the engine, making starting difficult and increasing the fire hazard.

These carburetors have a float, not just a gasoline pickup tube. The purpose of the float is to keep a constant level of fuel in the carburetor body regardless of the level of fuel in the tank or of the varying pressure created by the fuel pump. The pump or, as the case may be, gravity, feeds gasoline into the float chamber. As the chamber fills, the float rises and moves a needle valve, which stops more fuel from entering the chamber. The engine is, of course, using fuel, and as the level in the chamber drops, the float sinks and allows the needle valve to open. This fuel-metering cycle is repeated as often as 200 times a minute.

Floats may be simple cannisters with a needle fixed into position on the top as is European motorcycle practice, or may

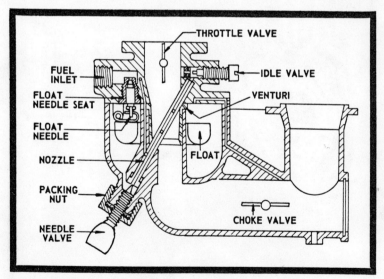

Fig. 3-7. Series D Lawnboy Sidedraft Carburetor. This design includes a primer for quick starting. (Courtesy Gale Products.)

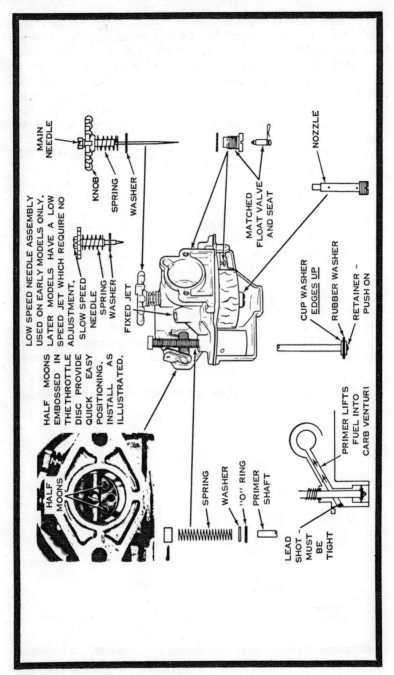

Fig. 3-8. Typical Updraft Carburetor. (Courtesy Briggs and Stratton.)

work the needle through a pivot hinge and tang arrangement. The latter is typical of American and Japanese design. Some float chambers are mounted remotely and feed more than one carburetor.

Light materials such as varnished cork, plastic, or sheet brass are commonly used materials for floats. Brass floats can develop leaks from corrosion or from vibration. One test is to place the float in a container of hot water. The heat will open any pin holes and the float will sink. To repair, heat the float to evaporate all the liquid and apply a dab of solder. Be sparing with the solder, because added weight will throw off the float adjustment. Treat these floats with care; they are quite fragile and can be damaged easily. The practice of indiscriminately "blowing a carburetor out" with high-pressure air can collapse the float. This can be quite embarrassing, especially if the customer happens to be watching.

As previously mentioned, American floats have a tang and pivot hinge arrangement to move the needle. The height of the float is quite critical, since this controls the amount of fuel in the chamber. Too much fuel will cause flooding and rich running, while too little will cause hard starting and lean high-speed operation. To adjust the float, assemble the inlet needle and the float and turn the carburetor body upside down. Measure the height of the float as shown in Fig. 3-9. This measurement will vary with different designs, but as rule of thumb for small American industrial engines, the float should be level with the cover. If you are in dobut about the adjustment, contact the parts distributor. Some carburetor rebuilding kits contain carboard templates which simplify the adjustment. Outboards are often equipped with a brass plug in the side of the float chamber. This plug can be unscrewed to show the actual fuel level. When the float is set correctly, fuel will just wet the threads when the carburetor is horizontal to the ground. Never press on the float to bend the tang, as this might damage the inlet needle.

Sometimes the carburetor will continue to flood after the float has been checked and set. The problem, by elimination, must be with the needle and seat. In years past, needles were constructed of nickel (stainless) steel and the seats were of brass. These needles should be replaced if they have noticeable wear or if they have become sharply pointed. Some mechanics lap in steel needles and seats with valve grinding compound to make a short term fix, but this practice is frowned upon. "Shady tree" fixes are clever, but are no substitute for new parts. Plastic needles, or steel needles with synthetic tips, are becoming standard. These seem to wear

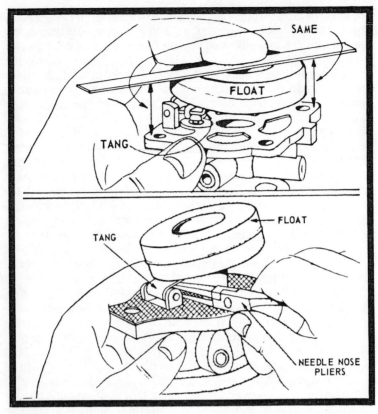

Fig. 3-9. Adjusting the Float Height. On this carburetor, as on many others, the float should be parallel to the body mounting surface. Do not press on the float to adjust. (Courtesy Briggs and Stratton.)

longer than steel and can tolerate small dirt particles without leaking.

American carburetors have a throttle plate (or "butterfly") which controls the amount of fuel-air mix going into the engine. The butterfly is mounted on a shaft which runs through the carburetor body. On the better designs, the shaft rides on bronze bushings. Eventually, the shaft will wear and allow air to leak into the engine. Sometimes this leak is so severe that the engine will refuse to idle. The shaft and bushings can be replaced, but on designs without bushings it may be necessary to replace the carburetor body. It is possible to purchase bushings which have an inside diameter matching the throttle shaft and then ream the carb body to accept them. The choice is between parts (a typical American carburetor body will cost anywhere from ten to twenty dollars) and labor.

European and Japanese carburetors usually employ a slide throttle rather than a butterfly. The advantage is that at full throttle there is only the width of the needle to restrict air moving through the venturi. On American designs, the throttle shaft makes a considerable restriction. Look at the Kawasaki design in Fig. 3-10: the throttle slide (No. 16) carries a needle (No. 38) which fits into the fuel nozzle (No. 39) and retracts as the throttle slide moves upward. As it retracts, more fuel passes through the nozzle and into the carburetor bore. The total amount of fuel is limited by the size of the main jet (No. 26). (In other carburetors, the main jet may be located in the nozzle.) The pilot screw (No. 19) controls the amount of air going to the low-speed side of the carburetor. Other designs, such as the Amal Monobloc shown in Fig. 3-11, have additional features. The throttle slide (376-060) has a cutaway which can be varied for special operation conditions. The slide also has a positive stop screw to limit how far down in the bore it can travel. This screw (shown in the inset as part number 376-070) thus controls the idle speed.

Cleaning the Float-Type Carburetor

Float type carburetors are complicated, and contain a large number of small ports and passage ways which must be cleaned prior to tuning. The procedure is to remove the carb from the engine and strip it completely. Be careful to use screwdrivers with the blades ground to exactly match the slots in the jets. The slots can be "jimmied" and throw off the mixture. If you are disassembling an unfamiliar design, lay the parts out on a clean bench in the order that they were removed. On Clinton, Tecumseh, Lawson, and Lawnboy engines the main jet nozzle (see Fig. 3-7) may have a tiny hole cut through on the side of the threaded end. This hole is not much larger than a human hair, but it is critical for low-speed operation. It must be aligned with a port in the carburetor body. You can uncover this port by removing a small brass expansion plug which is on the underside of the carb body, near the top of the float chamber. Use a small screw extractor and be careful not to lose the plug. Now, when installing the nozzle, tighten it until you can insert a piece of stiff wire (a wire wheel is a good source) through the port and into the hole. This operation takes a little patience, but it is routinely done in some shops. Replace the plug. An alternative is to purchase a replacement nozzle, which has a slot cut around the threads so that gas will flow regardless of alignment. These nozzles cost about a dollar each.

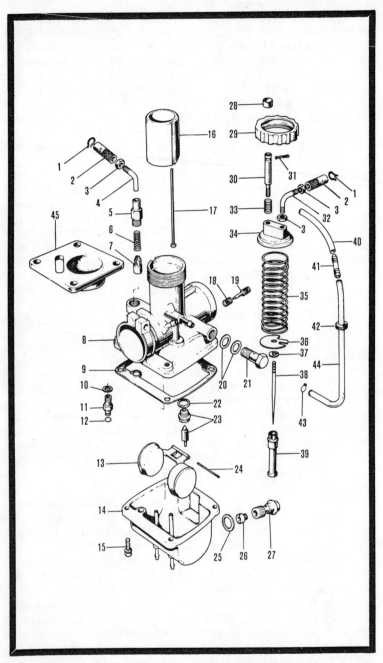

Fig. 3-10. Slide Throttle Carburetor used on the Kawasaki Model 175 F7 Motorcycle. (Courtesy Kawasaki Heavy Industries, Ltd.)

Sometimes the carburetor body will be so corroded that it will be difficult to remove the fittings. Penetrating oil helps and with the proper taps you can chase the threads up to the fittings. The rounded end of the tap should be ground flat so that the tool will cut all the way to the fitting.

Separate all plastic and composition gaskets and soak the metal parts in cleaner. Generally four hours should be the limit, because these cleaners will eventually attack the pot metal, discoloring it. Remove the parts from the solvent and blow dry. Inspect the carburetor mounting flanges to see that they have not been distorted. "Rabbit-eared" flanges can be ground with emergy cloth in the same way that cylinder heads are. Float chamber covers can be trued with light hammer taps on the mounting lugs.

Adjusting the Float-Type Carburetor

Now that the carburetor is thoroughly cleaned, it can be assembled on the engine and adjusted. The American butterfly designs will be discussed first:

1. Set both the high- and low-speed needles one and a half turns open.
2. With clean, fresh gasoline in the tank, run the engine to operating temperature with the air filter in place.
3. Adjust the high-speed needle first. The needle may be below the float bowl or on top of the carburetor body (Fig. 3-6) or on the side of the body, in line with the low-speed needle. If you are not certain, look for the low-speed needle—it will always be the one nearest the butterfly. The high-speed jet should be adjusted to give smooth response when the throttle is blipped. "Four-stroking" (firing every other revolution) on two-cycles is a sign of an over-rich mixture. Black smoke indicates the same thing on four-cycles.
4. Adjust the low-speed needle, until the engine idles at the highest rpm.
5. Check the high-speed adjustment, since the low- and high-speed circuits are somewhat interrelated.
6. Adjust the idle stop screw to give the desired idle speed.
7. If possible, test under load.

Slide throttle carburetors are somewhat more difficult to adjust, but most of the frustration can be avoided by approaching the job systematically. The high-speed side can be adjusted using two items; the position of the needle relative to the throttle slide, and the size of the main jet. If you look again at the drawing of the Kawasaki carburetor (Fig. 3-10) you can see that the needle (No. 38) has grooves cut into it. A spring clip (No. 37) fits on one of the grooves and locates the needle.

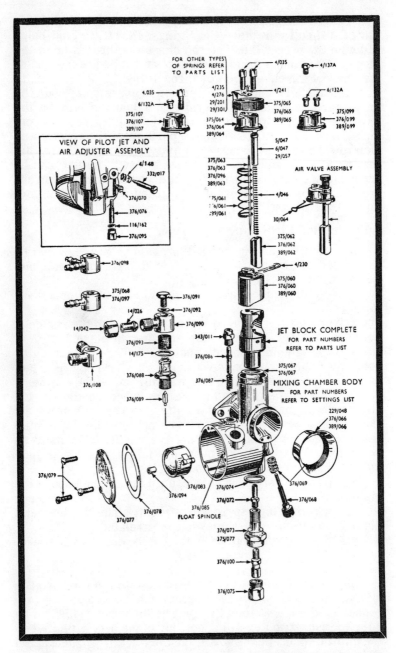

Fig. 3-11. The Amal Monobloc carburetor in Exploded View, and with Parts Numbers. (Courtesy Johnson Motors, Distributors for Triumph motorcycles.)

The higher the needle is raised, the richer the mixture. The Amal, shown in Fig. 3-11, uses the same system while other makes, such as the Villiers, have a threaded needle that can be moved in and out of the slide. The main jet on all of these carburetors is replaceable and comes in various sizes. Each jet will have a number stamped on it, which is the factory code for the inside diameter. Generally, the higher the number, the larger the jet.

The low-speed side has an adjustable needle working just like the familiar American butterfly design. However, the needle on Amal type carburetors controls air—not fuel—and adjustment procedure is the reverse of the American. As the needle is turned outward (counterclockwise) more air enters the low-speed circuit and the engine runs leaner. As it is tightened, the air flow is reduced. As a general rule, air control needles have a blunt end, while fuel needles are pointed.

Another adjustment is available on some of these carburetors. Throttle slides can be purchased with different sized cut-aways. The larger the cut-away, the more air will flow through the carburetor at any given throttle setting up to three quarters open. Past this point, the cut-away is overridden by the venturi cast in the carburetor bore.

Adjustment is as follows:

1. Set the main jet needle in the middle position relative to the slide.

2. Set the air screw about three quarters of a turn open.

3. With clean, fresh gasoline in the tank, run the engine a few minutes to reach operating temperature with the air filter in place. Do not run motorcycles more than 5 minutes without some method of forced draft cooling.

4. From three-quarters to full throttle, the mixture is controlled by the size of the main jet. Generally, the factory-supplied jet will be correct, although high-altitude operation or extensive modifications to other parts of the engine might dictate a change. An overly rich mixture will show these symptoms:

    a. A bluish white exhaust flame which may smoke.

    b. A blackened sparkplug (see the chapter on ignition systems for information on "reading" spark plugs),

    c. Performance is sluggish,

    d. The engine may run faster at part throttle than at full throttle. Make this test under load to protect the conn rod.

A lean mixture, i.e., one that is produced by too small a main jet, will give the following symptoms:

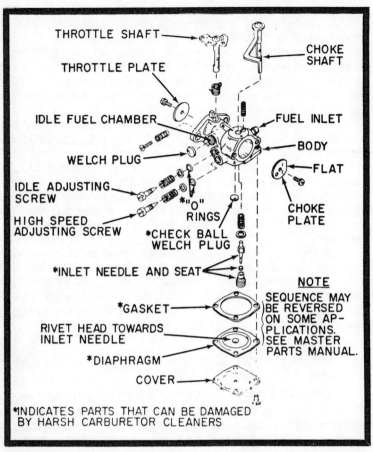

Fig. 3-12. Tecumseh Single-Diaphragm Type Carburetor. When assembling be careful to position the diaphragm as shown in the drawing—the rivet head must be towards the needle. (Courtesy Tecumseh Products Co.)

    a. The engine runs hot,
    b. The spark plug tip is white.
    c. Engine rpm fluctuates with the throttle held steady.

  5. From one-quarter to three-quarters throttle, the mixture is controlled by two factors: the size of the cutaway and the position of the main jet needle in the slide. Few mechanics ever change the cutaway, and only then on racing or other special-purpose machines. It is much easier to move the needle up for a richer mixture and down for a leaner one.

  6. From idle to one-quarter throttle, the mixture is controlled by the low-speed (air) needle. Find the position where

the engine runs at highest rpm with the throttle set. It may be necessary to adjust the throttle valve stop screw to get the desired idle rpm.

7. Test the machine under load.

Diaphragm-Type Carburetors

If the present trend continues, diaphragm carburetors will displace the float types, at least as far as American engines are concerned. There are a number of reasons for this popularity: the diaphragm carburetor is usually more efficient than the other types; it is cheaper to manufacture; and many of them incorporate a built-in fuel pump which, along with the absence of a float, means that the engine can operate at any angle off the vertical.

These carburetors operate by means of variations of crankcase pressure. As the piston moves toward the combustion chamber, it leaves a low pressure area behind it; and as it returns on the downstroke it compresses the air (or the air and fuel mixture in the case of two-cycles) in the crankcase. These fluctuations in crankcase pressure are transmitted by means of a port drilled in the intake pipe.

Two types of diaphragm carburetors presently see service. One type, now being phased out, is simply a carburetor with a diaphragm doing the work of the float, that is, the diaphragm controls the amount of fuel delivered to the main and pilot jets. An exploded view of this type is illustrated in Fig. 3-12. The underside of the diaphragm is vented to the atmosphere. Low crankcase pressure causes the diaphragm to lift and move the inlet needle off its seat. A tiny amount of gasoline then flows around the needle and into the fuel chamber.

The second type of diaphragm carburetor is very similar to the above, but has the additional refinement of a fuel pump. Either there will be two separate diaphragms in the separate chambers above each other, or more commonly, there will be a single diaphragm and a fuel pump element as in the Tecumesh pattern shown in Fig. 3-13. On inflation (Fig. 3-14) the pump element (1) expands and forces fuel out of the pump cavity (2) against the inlet fitting check valve (4) which closes, and into the needle valve by way of the body check valve (5). From this point, the carburetor functions as described in the preceding paragraph.

On deflation of the pump element (Fig. 3-15), fuel is drawn through the inlet fitting check valve (4), and into the pump cavity (2). The check valves usually consist of plastic discs

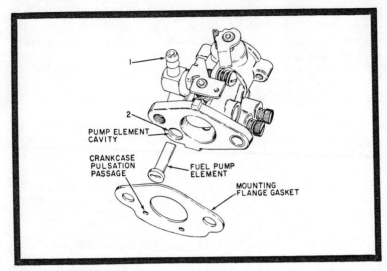

Fig. 3-13. Tecumseh Diaphragm Type Carburetor with a Built-In Fuel Pump.

although some models employ a nylon ball in the inlet fuel fitting. In any case, caustic cleaning agents, such as Bendix, should be avoided. Varsol, or in extreme cases, a brief immersion in lacquer thinner, is all that can be recommended.

Another interesting design of a diaphragm type carb is shown in Fig. 3-16. This carburetor has an accelerating pump which delivers extra fuel to the venturi as the throttle is opened. The main jet is fixed (although different sizes are available). To adjust this carburetor, close the idle and the intermediate-speed needles and open ⅞'s of a turn. Start the engine and allow it to reach operating temperature. Adjust the intermediate speed needle to the setting which gives the highest rpm. Richen (open) it another ⅛ turn. Adjust the low speed and the idle stop screw for the desired idle (900 to 1100 rpm) and re-check the intermediate setting.

The inlet check valves can be removed for service by carefully twisting and pulling the inlet fitting out of the carburetor body. On replacement, the fitting should be coated with a small amount of Loctite and pressed home. The carb body check valve is also removable, but requires a puller. Normally, this repair is not made, since labor costs exceed the price of a new carburetor body.

All valves and the fuel pump element are tested by blowing against them. The use of compressed air is destructive to these delicate parts.

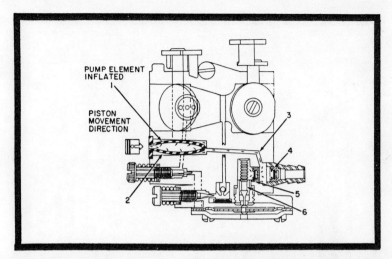

Fig. 3-14. Diaphragm Operation. High crankcase pressure expands the fuel pump element, forcing fuel past the body check valve (5) where it collects at the needle valve (6). (Courtesy Tecumseh Products Co.)

In spite of their performance capabilities, these diaphragm carburetors generally have a poor reputation among mechanics. More of them seem to be in the shop than their numbers warrant. The complaint is always the same: hard starting.

The major problem is the diaphragm. If it stretches, the engine will be hard to start, although it might run well once

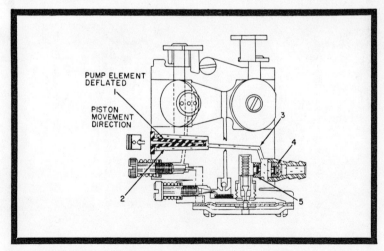

Fig. 3-15. The fuel pump element is deflated due to low pressure in the crankcase. The carburetor is now "charged" with gasoline in the fuel pump cavity. (Courtesy Tecumseh Products Co.)

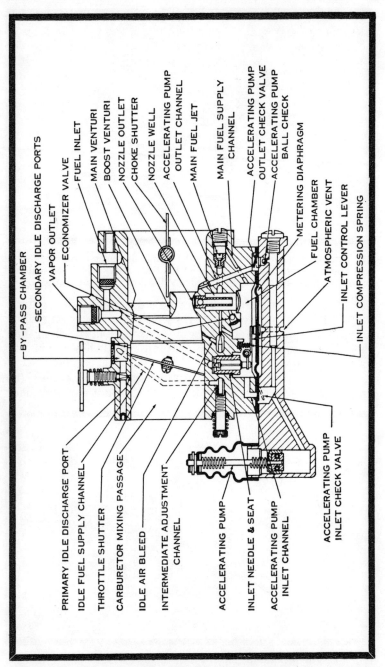

Fig. 3-16. The HD Diaphragm Carburetor with Accelerating Pump, used on the 1970 Harley-Davidson Electra Glide. (Courtesy Harley-Davidson.)

started. Stretching can be caused by age or by stale fuel left in the system. To replace the diaphragm, remove the four X-headed screws which hold the cover to the bottom of the carburetor body. On rotary lawnmowers, the usual procedure is to remove the entire carburetor from the engine, but the diaphragm cover plate can be removed with the carburetor in place if you use an off-set ratchet screw driver. Be sure the outer surfaces of the carb are clean before you disassemble it. A new diaphragm is installed as per instructions on the package, and the cover is replaced.

Cleaning the Diaphragm-Type Carburetor

If the engine still refuses to start, and you are sure that it is a fuel problem, disassemble the carburetor completely and clean it, observing caution with the plastic parts. Look carefully at the inlet needle and seat. If the needle is bent or if the seat is scored, the carburetor will "leak down" during cranking. The main and idle jets are sometimes found to be plugged by broken tips of the adjusting needles. These needles are made of soft iron and fragment against the jet if they are over tightened. On the fuel pump types, replace the element and alternately blow and suck on the check valves to see that they are not leaking.

Adjusting the Diaphragm-Type Carburetor

Tuning is not difficult if approached systematically:
1. Fill the tank with clean fuel.
2. See that the choke closes completely and without binding. Lawnmowers and other small, single-cylinder engines will be difficult to start if the choke does not fully close. The choke plate has a cut-out on it to provide enough air for combustion.
3. Set both the idle and the high-speed needles open at about three-quarters turn. This should provide enough fuel so that the engine will run.
4. With the air filter in place, run the engine for 5 minutes to reach operating temperature.
5. With the choke open, adjust the high-speed jet. This will be the jet farthest away from the throttle plate. Black smoke and exhaust smelling strongly of gasoline indicate an over-rich mixture. Blip the throttle, and richen if the engine stumbles.
6. Adjust the low-speed needle for the highest rpm at idle.
7. Reset the high-speed, and check the low-speed again.

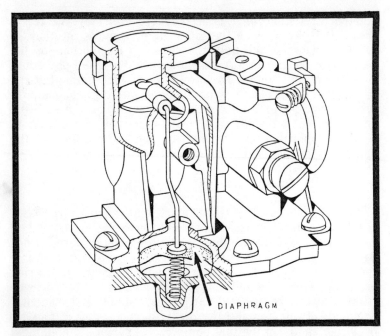

Fig.3-17. A Carburetor with Vacuum Actuated Choke, used widely on Briggs and Stratton engines. Other manufacturers employ automotive type checks which respond to manifold heat. (Courtesy Briggs and Stratton.)

8. If possible test under load. Often you will find that the high speed side needs to be slightly richer.

Briggs Composite Carburetors

Briggs manufactures three carburetors known together as the "Pulsa-Jet" series, which combine a diaphragm fuel pump with a suction lift tube. You may ask, "Why have a fuel pump?" The answer is that suction lift carburetors tend to run lean when there is little fuel in the tank, and overly rich when the tank is full. This is why I specified before that the tank be half-full whenever the carburetor is adjusted. The Pulsa-Jet pump (Fig. 3-17) delivers fuel to a reservoir built into the top half of the fuel tank. The suction tube picks up gasoline from this reservoir, which is always at a constant level. If the pump delivers more fuel than the engine can use, the surplus goes back into the tank through an overflow.

Like suction-lift designs, Pulsa-Jet carburetors are mounted directly on the top of the tank. Two tubes extend out of the bottom on the carburetor body into the tank. The longer tube extends almost to the bottom of the tank. Fuel is picked up in this tube and is expelled through a port in the carburetor body into the reservoir. The shorter tube picks up fuel from the reservoir and delivers it to the main and low-speed jets.

These carburetors present few problems, other than the large number of plastic parts used, which makes complete disassembly necessary prior to chemical cleaning. The diaphragms are easily accessible and should be replaced as part of every tune-up. Pickup tubes with hexagonal heads are meant to be unscrewed from the carburetor body. The round-headed tubes are pressed into place. Pulsa-Jet pickup tubes do not have ball check valves, and replacement is only necessary if the tube has been damaged or if the screen is clogged beyond the possibliity of cleaning. Brass tubes are **not** designed to be removed. If the screen on a brass tube must be replaced, secure the tube in a vise (being careful not to crush it) and pry the screen assembly off.

Multiple Carburetors

Automobiles generally have a single carburetor which feeds as many as eight cylinders, through the intake manifold. Small engine designers prefer to use one carburetor per cylinder. Thus, four-cylinder Hondas have four separate Kehin carburetors. There are several good reasons for this complexity: first, it is difficult to design the perfect manifold. Almost inevitably the cylinders remote from the carburetor will run slightly lean and those near the carburetor will be rich. This means a loss of power and of fuel economy, since the mechanic must settle for a compromise setting. Another problem is that no two cylinders are exactly alike—each has its own peculiarities which can only be compensated for by individual carburetors adjusted to the demands of each cylinder. And finally there is the matter of sales appeal. A four-carburetor Honda is somehow more sporting than one with a single pot, even though the single venturi would have to be as large as the combined area of the four.

To adjust multiple carburetors, the guiding principle is to work on one at a time. Get one right, and then move to the next. Make sure that the throttle cables on motorcycles or the throttle levers on outboards or snowmobiles move all throttles simultaneously. If the throttle movement is not synchronized, the individual cylinders will want to run at different speeds. Now, warm up the engine and disconnect the spark plug wires to one or more cylinders. Adjust the carburetor on a running cylinder as indicated previously. Disconnect that cylinder and go on to the next. Suspect trouble if the adjustments on one carburetor vary widely (a turn or more) from the others. This kind of variation might mean that a jet is clogged or that there has been a manufacturing error. Connect the spark plug leads to all cylinders and test. If the idle speed is too high, it can be lowered by turning the throttle stop screw an equal amount on all carburetors.

If you have occasion to remove the carburetors from the engine, note that they may not be identical. Some manufacturers (such as Suzuki) specify a left and a right carburetor; they are distinguished by the position of the adjustment needles on the carb body.

Automatic Chokes

The manual choke rarely gives trouble. But it does require some attention from the user; the choke must be closed completely for cold starting on most engines. As the engine loses compression with age and wear, the choke setting can be critical even when the engine is warm. We all know of "dogs" which have a secret combination of choke and throttle settings before they will run. And if the choke is left partially closed on a running engine, problems rapidly develop. Two-cycles foul plugs and ports, and four-cycles suffer rapid upper cylinder wear since the surplus gasoline washes the oil off the bore and out the tail pipe. An automatic choke which compensates for engine heat and outside temperature is desirable.

Since August of 1968, Briggs and Stratton has used an automatic choke on their smaller engines. This choke does not adjust to temperature changes, but it does insure that the engine is fully choked on starting and "un-choked" when running. The mechanism works by means of engine vacuum: when the engine is not running, there is no vacuum and the spring-loaded choke plate closes, and as the engine starts, the plate is pulled open. This is an admirable solution to the problem, but it does call for some special service procedures. Refer again to Fig. 3-17 which shows a typical Pulsa-Jet carburetor and a cutaway of the choke mechanism. The top of the tank and the carburetor body must be airtight. If you suspect that the tank is warped, lay a machinist's straight edge across it at several points. A .002" feeler gauge should not slip under the tank and the straight edge. The critical areas are around the fuel ports and channels, and around the vacuum ports. A warped tank should be replaced, although short term repairs are sometimes made with silicone cement. Be careful to keep the cement out of the ports. Normally, the carburetor body will not be warped.

Repair and Replacement

The choke diaphragm should be routinely replaced. On assembly, pre-load the diaphragm by inserting a ⅜" diameter bolt into the air horn (Fig. 3-18) in such a way as to hold the choke plate open. This will stretch the diaphragm. Next, tighten the carburetor mounting screws evenly and in a criss-cross fashion. Remove the bolt and test. The choke should close when the engine is not running, and should open fully once the engine starts.

Kohler and other manufacturers of larger industrial engines use thermostatically-controlled chokes. The heart of the mechanism is a bimetallic strip coiled in the form of a spring, which is heated as the engine runs. The strip consists of two dissimilar metals, bonded back to back. Usually one is copper and the other is iron. As the strip gets hot, the copper expands more than the iron, and the strip tends to uncoil. One end of the coil is anchored to the housing, and the other end is connected to the choke plate. Another mechanism consists of a small piston in the carburetor which responds to engine vacuum. The piston is connected to the choke plate and, once the engine starts, pulls the plate open. However, the bimetallic

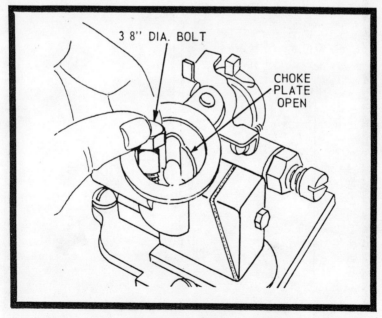

Fig. 3-18. Pre-setting the Choke Diaphragm on a Pulsa-Jet Carburetor. (Courtesy Briggs and Stratton Corporation.)

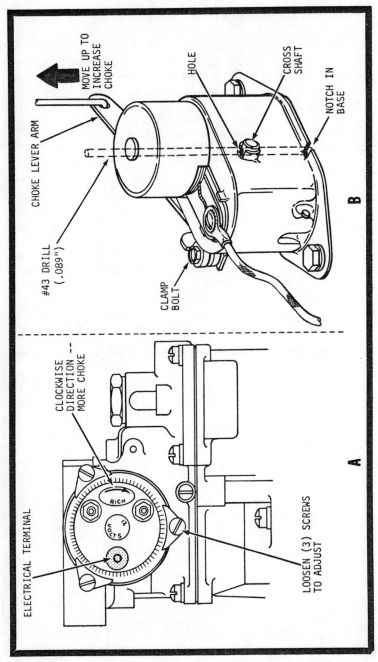

Fig. 3-19. Two styles of Automatic Chokes used on Kohler Twin-Cylinder Engines. (Courtesy Kohler of Kohler.)

coil tends to keep the choke closed. There is a balance of forces here. Eventually, as the engine heats up, the coil will unwind, and the piston will pull the plate open.

Some choke coils are built into the side of the carburetor, while others are mounted remotely and move the plate by means of a link. (Fig. 3-19 shows two Kohler designs.) The source of heat may be the exhaust manifold, or an electric heating coil.

Testing

To test the choke, remove the air filter and manually move the plate. On a cold engine, it should be nearly closed. (The exact setting varies from carburetor to carburetor and can be found under the manufacturer's table of specifications, although, normally, small engine mechanics do not tamper with the settings.) Flip the plate with your finger. It should move easily and return to its initial closed position. Start the engine. In a few minutes, the choke should open completely. On some designs, it may be necessary to blip the throttle to unlatch the choke mechanism.

Various things can go wrong with these chokes. The remote manifold heat type have a tube or a hose going from the manifold to the bimetallic coil. This tube can become clogged with carbon. With electrical chokes, the current should flow when the ignition is turned "on." Usually, the ignition switch contains a separate element for the choke, which may fail, even though the ignition coil portion of the switch functions. Another problem is binding of the choke mechanism, especially around the vacuum piston and the choke plate shaft. Automatic choke cleaner, sold in aerosol cans, may be useful, but a surer repair is to disassemble the carburetor and soak it in Bendix solution. And a weak bimetallic coil can cause the choke to refuse to close. The best solution is to replace the coil with a new one.

The choke shown in Fig. 3-19A is adjusted by loosening the three mounting screws and rotating the choke body. Turning it in the direction marked "Rich" tightens the coil and keeps the engine choked longer. This adjustment is helpful in cold weather operation. The remote type shown in Fig. 3-19B is set by loosening the clamp bolt on the choke level. A No. 43 drill bit (.089") is then inserted through the cross shaft hole and engaged in a notch in the base. The dotted lines on the illustration show the position of the drill bit. Move the plate to the closed position, tighten the clamp bolt, and remove the drill bit.

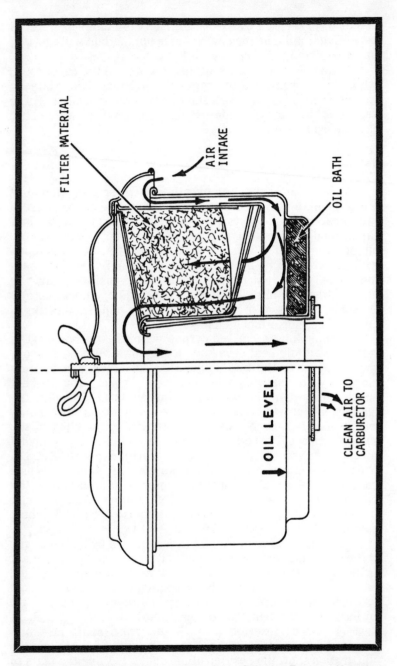

Fig. 3-20. Oil Bath filter used on Kohler Twin-Cylinder Engines. Note that these filters are not used on maximum performance engines. (Courtesy Kohler of Kohler.)

Air Filters

The air filter has several functions on modern engines. The primary function is, of course, to trap dust particles before they enter the engine. Filters also help contain carburetor fires—an especially important function in marine applications—and recently have been designed with the view of reducing intake tract noise. No engine, and particularly no two-cycle, should be operated without a filter.

The earliest type air filter consists of a "dough nut" of pressed steel or aluminum shavings between two metal discs. The assembly is bolted directly to the carburetor body. Air must pass through the outer circumference of the "doughnut," and, presumably, the dust particles are trapped in the mesh of shavings. These filters may be cleaned with solvent, and should be oiled to increase their efficiency. Many small two-cycles still use this type of filter. These engines typically have a fog of oil-fuel mix which forms around the carb intake, thus keeping the filter wetted. The biggest problem with most of these is that the filter vibrates and eventually pulls the mounting bolt threads. This can be prevented by coating the bolts with Loctite prior to assembly.

An improvement over the dry mesh filter is the oil bath type. As shown in Fig. 3-20, these designs have their own oil supply which keeps the mesh wetted. Also, the incoming air must abruptly change direction. Dust particles, having some mass, continue in a straight line and become trapped in the oil. These filters should be routinely cleaned and the oil replaced. Some problems occur with the gasket between the filter and the carburetor. Overfilling the reservoir or tilting the engine much beyond the horizontal will cause oil to enter the carburetor. Some dishonest mechanics boast of "rebuilding" an engine simply by draining the excess oil out of the filter.

Modern filters are much more efficient than the designs discussed above. Some manufacturers claim a filtration of particles as small as 25 microns in diameter (a micron is a millionth of an inch). Two types are currently supplied. The plastic foam or polyurethane elements (Fig. 3-21) can be cleaned and re-used indefinitely. Detergent and water or solvent is used to clean the filter and a small amount of motor oil—no more than a teaspoon for a single-cylinder engine—is kneaded into the element. Excessive oil is as bad as too little. Clinton goes so far as to suggest that no oil be used if a petroleum-based solvent is used for cleaning. The big disadvantage of these filters is that they can be destroyed by flame. Backfiring will soon ruin one.

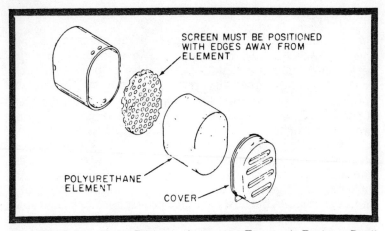

Fig. 3-21. Polyurethane Filter used on many Tecumseh Engines. Don't over-oil these filters, because too much oil will cause the engine to run rich. Clinton suggests that their filter be washed in kerosene and that no oil be used. (Courtesy Tecumseh Products Co.)

The second type of modern filter is paper (Fig. 3-22), rolled in the form of a cylinder and pleated to increase its area. These filters must **not** be cleaned or oiled, since liquids cause the fibers to swell closed. The filters are usually well protected from rain, but an over-enthusiastic application of a water hose, as when washing a motorcycle, can ruin them. On disassembly, the filter should be checked for cracks. Tap it on a hard surface to remove accumulations of dust. Some mechanics blow them out with low pressure air. Paper filters are replaced as per the manufacturer's suggested schedule, but a rough and ready test is to see if the filter will pass light. Illuminated by a flashlight in a darkened room, the filter should glow like a Jack O'Lantern.

FUEL TANKS, LEAKS, AND REPAIR

Made of aluminum, steel, and (unfortunately) plastic, fuel tanks come in various styles and sizes. Tanks fitted to vehicles and boats should have internal baffles to reduce the instability caused by "free surface." (The effect of free surface can be seen by a little experiment with an empty ice tray. Remove the partitions and fill the tray with water. Now do the same thing with the partitions in place. The partitions improve the balance of the tray by cutting into the free surface area.) Check that the internal baffles haven't collapsed.

The tank should be inspected for leaks whenever the engine is serviced. Consider the potential stored in even a small tank: a quart of gasoline has the explosive energy of approximately two pounds of dynamite. Small engines are usually designed so that the tank is mounted above the exhaust pipe. Leaks from the tank or fuel lines collect around the exhaust, are ignited, flame back to their source, and explode.

Most leaks on small industrial engines originate at the fuel cap. The gaskets on the cap shrink with age, and engine vibration "walks" fuel up and out of the spout. The cure is to replace the cap, or if the spout has been damaged by overtightening, to replace the tank as well. Engines with suction-lift carburetors are especially dangerous. The exhaust exits directly over the tank. Fortunately, motorcycle and outboard fuel tank caps rarely leak.

Other leaks can be repaired, although most mechanics prefer not to bother with the job. The safe way is to use a good grade of epoxy cement (such as that sold by Whirlpool distributors). Thoroughly clean the area to be patched with sandpaper and lacquer thinner and apply the epoxy, mixed as instructed on the tube. Give the cement 24 hours to cure and then test it.

Another method is to solder the tank. This method is reliable, but can be dangerous if you do not take the proper precautions. First steam-clean the tank. If you do not have the equipment, take it to a radiator shop and have it done. Gasoline fumes linger for years in an "empty" tank. Use acid-

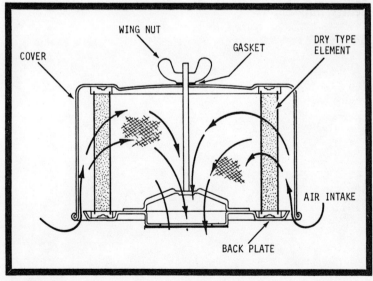

Fig. 3-22. Paper-Element Filter. These filters must be kept dry and must be replaced regularly. (Courtesy Kohler of Kohler.)

core solder and a gun—never solder a tank with an open flame.

Leaks on the gas tank proper will be around its mounting bolts or on the edges where it is crimped together. Water in the gasoline will collect at the bottom of the tank and rust out the lower portion. This kind of damage is not repairable. Once the tank begins to rust, it should be replaced.

Small tanks are usually dumped into the cleaner and repainted afterwards. (Distributors carry original factory colors in spray cans and matching decals.) Motorcycle tanks can be cleaned without marring the finish if you are careful. Cleaning can be speeded by dropping a few large brass bolts into the tank and agitating. Brass or some other nonferrous metal should be used to preclude sparks.

Sometimes it will be necessary to clean a tank. Evaporated fuel will leave a heavy, black varnish which can only be removed chemically. There are various preparations on the market for this, but most mechanics seem to prefer Bendix Econo-Clene. It comes in five-gallon tins with a built-in tray for small parts. For small jobs you can purchase Gunk Carburetor Cleaner in pints. CAUTION: These chemicals are potent, and will remove paint and skin. Of course, do not use any of these cleaners on plastic tanks or on Briggs and Stratton steel tanks. These latter are unusual since they are crimped together and sealed with a plastic filler. After a half hour or so in the cleaner, the filler would soften. Repair is nearly impossible.

Chapter 4

Electrical Systems

Most components of small engines have not changed much over the years; pistons, rings, and even carburetors are pretty much what they were a generation ago. The big change has been in the electrical systems. Two factors are at work here: The customer has demanded more convenience—electrical starting and, in many cases, a surplus of power to operate lights and other accessories—and breakthroughs in technology have given rise to new concepts in the generation and distribution of electric current.

ELECTRICAL THEORY

An electric circuit is an uninterrupted path through which electrons flow from the negative (-) terminal of the voltage source (a battery or a generator), through a load (spark plugs, lamps, for example), and back to the positive (+) terminal. A circuit must have three components: a voltage source, a load, and the connecting wires. Two-wire direct-current circuits use a separate wire for the negative and the positive, and single-wire circuits use one "hot" wire and the frame of the machine for a ground. Most of the circuits you will encounter in small engines are the single-wire variety. Two-wire circuits are found on some Japanese motorcycles and occasionally on power boats. The advantage of the two-wire arrangement is the relative freedom from shorts (since both conductors must be bare and touching) and from fire hazard. The drawback is the expense and the bulk of the extra wire used.

If there is only one possible route for electrons from the negative terminal and back to the positive terminal, the circuit is called a series circuit. In a simple ignition circuit, the ballast, starter solenoid, and ignition switch are connected in series with the battery. When the switch is open, no current can flow anywhere in the circuit. Years ago, Christmas tree lamps were wired this way. If a single lamp burned out, all the lamps on the string went out since no current could flow through any of them.

If there are two or more paths supplied by a common voltage source, the circuit is described as parallel. The coils and points on the Scott are in parallel. One side is connected to the hot wire and the other side goes to the ground. Thus current can flow through either or both contacts. Ordinary house wiring is parallel, with a fuse to protect each branch circuit.

Small engine wiring is a series-parallel combination. Switches, main line fuses (as in the Honda), and rectifiers, are put in series with the voltage source, while lights, horns, and starter motors are wired in parallel. However complicated the circuit might appear, it will always consist of simple series and parallel wiring.

Troubleshooting

A dual outboard installation, fitted with remote controls and with deluxe accessories, may involve dozens of separate circuits and a thousand feet of conductor. Yet a good technician can isolate a malfunction in a matter of minutes. He does this by working systematically from a few general principles.

An electrical component can fail in one of two ways: it can refuse to pass current or it can short. When a component refuses to pass current, we say that it is open. A short means that the current flows, but around the load and not through it. Shorts usually go to ground in single-wire systems. A bare conductor touches the frame and the electrons bypass the load. This can occur on two-wire systems as well, but it is comparatively rare since uninsulated portions of both conductors must be in physical contact. Another kind of short can occur in transformers, solenoids, field windings, and armatures. These components are built up of a large number of turns of fine wire layed over a form. Current is introduced at one end of the wire, travels through each winding, and exits at the other end. Along the way, it produces useful work in the form of magnetism. The turns of wire are wound one on top of the other, and are insulated with varnish or enamel. Plastic or rubber and cloth insulation would be electrically stronger, but would take up too much space. It is possible for the insulation to break down and for the windings to short out to each other. Instead of the current passing through each coil, it can then pass across the coils.

When checking for opens, inspect all terminals for cleanliness and for a tight mechanical fit. The ground connections are as critical as the "hot" side. Corrosion, grease, and paint can easily stop 6 or 12 volts. Pay particular attention to battery terminals and friction type connectors. Switches (including solenoids and relays) should be viewed with suspicion. Whenever the switch is opened, there is a tendency for the current to momentarily arc. Arcing encourages the formation of corrosion. Also, over a long time, arcing will destroy the contacts by metal transfer. Many switches are sealed, but temporary repairs can sometimes be made by flushing the switch with television tuner cleaner. Another source of opens is wear on the brushes of DC generators and starter motors. The brushes must be in solid, physical contact with the commutator. And opens can occur inside of an insulated wire. Fortunately, this condition is rare and is usually the result of physical damage. The copper strands will be severed, but the insulation may not be obviously damaged. A partial open connection—one which has high resistance and so only passes a small amount of current—will be warm to the touch.

Shorts also betray themselves by heat. Usually the heat and the damage will be more extensive than in a high resistance connection. Look for melted and charred insulation and discolored terminals. Shorts across coils can be detected by carbonized varnish. In extreme cases, the varnish will have boiled and pulled away from the wires, exposing bare copper. Composition board, such as used on fuse boxes and on lamp sockets, can become conductive by absorbing moisture. Heating the component in an oven at 250 degrees F. for a few hours is enough to drive the moisture out. The component then should be coated with epoxy to make a permanent repair.

Test Equipment

The tool used in troubleshooting is the volt-ohm meter (also known as a VOM or a multimeter), available from radio supply houses. These devices measure volts, amperes, and ohms. A rotary switch is used to select functions and to provide a range for each function. Prices vary from as high as $600 for the digital types to as low as $10 for some of the Japanese imports. A good average price is around $80, but many mechanics purchase a $20 multimeter and use it for years. The more expensive units are highly accurate and may have special features such as circuits to prevent burn-out and also to protect against reversed polarity. The Simpson 255 features a thermocouple which indicates temperature range of 100 to 1050 degrees F. Some shops employ these meters to determine coolant and cylinder head temperatures.

The heart of the typical volt-ohm meter is a D'Arsonval movement. This is a U-shaped permanent magnet with a coil suspended between the poles. The coil is free to rotate and has a pointer affixed to it. As current flows through the coil, magnetic lines of force developed around the coil react with the permanent magnet. The coil and pointer rotate in direct proportion to the amount of current flowing. Tiny balance springs return the mechanism to the zero setting when no current flow is present.

However, very little current is required to move the pointer to full scale position. If too much current is impressed on the coil, the pointer will peg against the stop, and windings will overheat and burn. To increase the range of the meter and to protect it from overloads, series or paralleled resistors are switched into the circuit. These resistors are cylinders of carbon or coils of wire wound over a ceramic form. Each resistor is carefully calibrated to allow only a preset amount of current into the coil. When measuring amperage, the resistors are placed parallel to the coil so that the excess current is shunted around it. For voltage measurements, the resistors are in series with the coil. The ohm function is performed by means of a battery in the meter case. The leads carry low voltage current to the test circuit. Resistance (R) is measured in ohms. Most multimeters have three ranges: R x 1, R x 100, and R x 10K ("K" stands for kilo or 1,000). To achieve these ranges on a single meter, resistors are plugged in and out of the circuit.

When testing for voltage or amperage, always set the meter to the highest expected range. If there is no reading, you can always switch down to a lower scale. Polarity is important in direct current measurements. Reversed polarity causes the movement to swing backwards, against the balance springs, and may cause permanent damage. Some multimeters have a diode to protect the movement, but most of the cheaper ones depend upon the operator's alertness. Before making any voltage or amperage readings, observe the polarity of the machine. On single-wire systems, one lead from the battery will go to ground. The other lead, called the hot side, will connect to the wiring harness. If the hot side is positive, readings are taken so that current flows through the positive (red) probe and out the negative probe. If the battery has a positive ground, the probes would be reversed. On two-wire systems, the best advice is to obtain the wiring diagram. (Many servicemen determine polarity by momentarily touching the leads to the circuit on the theory that a quick reading will not hurt anything.)

To read amperes, all the current flowing in the circuit must go through the meter. The circuit is broken and the meter connected in series with it. The circuit need not be broken to check voltage. Touch the bare conductor with one probe and ground the other to the frame or to the second wire. When measuring resistance, the test circuit must be disconnected from its voltage source, so that the only energy in the circuit will be from the battery in the meter case. Touch the probes together. The pointer should swing to zero, indicating no resistance. Actually there is some resistance in the leads, but multimeters are generally too insensitive to measure it. More than zero means that the battery has discharged. Some compensation can be made with the variable resistor on the meter case. With the leads apart, resistance should be infinite. The resistance of the portion of the circuit between the probes can be read out directly on the meter. An open will show as infinite resistance and a short will have zero or almost zero to ground. However, you should bear in mind that copper wiring has very low resistance per foot. A zero reading along its length is normal.

A multimeter gives the serviceman the option of checking the circuit for three functions. Any of these functions could be used to pin-point the trouble. But with the exception of adjusting regulators and bench testing starter motors, amperage is rarely measured, because the circuit must be broken and the meter inserted. It is much faster to detect the presence of current by running voltage checks to the ground or by making resistance checks. Most servicemen prefer the latter method because there is no danger of reversed polarities and meter burnout. Another advantage of using the ohm function is that components can be removed from the circuit and tested individually. A bad fuse or a lamp with a broken filament will show infinite resistance.

If you do much electrical work, the original test leads should be supplemented with a pair of Allied No. 732-6013, Type 8878 cables. These cables are equipped with jacks and test prods and while normally coiled, extend to five feet. An alligator clip should be soldered to one probe. This makes it possible to check spread out wiring systems such as found on power boats and on snowmobiles.

DIRECT CURRENT GENERATORS

An increasing number of small power plants are being equipped with battery-generator-starter systems. The primary purpose of the generator is to service the battery so

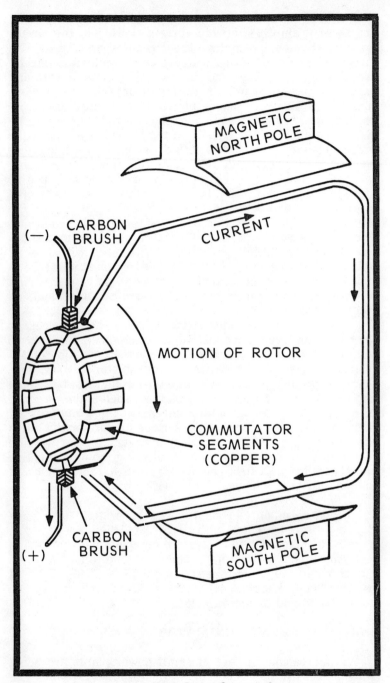

Fig. 4-1. Simplified Direct-Current Generator

that ample current will be available for starting. The generator is similar to the magneto in broad principles of operation (e.g., electric current is produced when magnetic lines of force are cut by a conductor), except that low voltage DC is produced.

Fig. 4-1 shows a two-pole generator. As the armature (the solid rectangular wire) revolves, it cuts the magnetic lines of force between the two poles, and current flows in the windings of the armature. Without the brush and commutator assembly, the current would alternate and be useless as a source of energy for the battery. The commutator is divided into as many as twenty-eight interconnected segments. Two soft carbon brushes are positioned 180 degrees apart so that they contact opposite pairs of segments. One brush is always negative and the other always positive. Depending upon the polarity of the system, either the negative or positive brush will be grounded. The "hot" side is connected to an insulated terminal marked variously "A," "Arm," or "D." All current comes off this terminal.

The field coils are electromagnets, and must have a source of energy. A generator produces current in proportion to the speed of rotation and to the amount of current going to the fields. Small generators are shunt wound, that is, the fields are connected in parallel with the armature. Most American and Japanese designs feed current to the fields from the "hot" side of the armature. This arrangement is known as type A. The British pattern (also found on American Ford cars) is the reverse: the fields are connected to the grounded brush. From the serviceman's point of view, the only important difference in these two designs is in the test procedures.

Construction

The typical generator as found on industrial and outboard motors has a frame (casing) made of cast iron. Besides being economical, this material is a good "conductor" of magnetic lines of force and so improves the efficiency of the generator. On motorcycles, a slight loss of electrical efficiency is sometimes traded off for lightness, and thus many of these generators are machined from aluminum. The fields are connected together in series and, as explained above, may or may not be grounded. Fields are wound around iron shoes or cores which are fastened to the frame by screws. Armature windings consist of lengths of varnished copper wire looped around a laminated iron core, and soldered to the commutator bars. Each bar or segment is insulated from adjacent segments and from the shaft by means of mica or some more modern material. Bearings are either oilite bushings or balls. Drive is by belt (on many American industrial engines and on the English Velocette motorcycle) or by gear.

Service

First it is necessary to determine if the generator, and not some other component, is at fault. Check the battery as explained later in this chapter, and disconnect the battery cables. With all switches "off," there should be infinite resistance between the cables. Low resistance means that there is a short somewhere in the wiring. The next step is to isolate the generator from the voltage regulator. On type A generators, the procedure is as follows:

1. With the battery in place, disconnect the "F" (field) terminal from the voltage regulator. If the engine is not equipped with a built-in ammeter, connect a meter in series between the "B" terminal and the battery. All current going to the battery will then register on the meter.

2. Run the engine at approximately 3,000 rpm. The generator should put out little or no current. A strong charge means that the fields have grounded themselves to the frame.

3. With all accessories switched "off," **momentarily** ground the lead from the field. If the generator is good, output will jump to maximum. Perform this test quickly to protect the generator.

On Lucas type B generators, the procedure is different. The fields are designed with internal grounding. Perform the following:

1. Run the engine as before and disconnect the field lead to the regulator as before. The generator should not charge.

2. With all accessories "off," **momentarily** connect the field lead to the armature terminal. Output should jump to maximum.

Once it has been established that the generator has failed, the next step is to determine exactly where the fault lies. Beginning with the obvious, examine the drive mechanism for signs of slippage. Typically, vee belt drives should have ⅜" free play between centers. On gear drive units, check for proper mesh and for broken teeth.

Most generator failures occur in the brush-commutator assembly. Inspect for chipping, cracking, and failure to make contact. The latter condition is caused by carbon dust in the brush holders, or by brushes which have worn to less than half of their new length. The commutator must be free of oil and dust. Television tuner cleaner, available from electronics supply houses, is excellent solvent.

For good brush life, the commutator segments must be smooth and true. Signs of arcing on a single segment usually means that its winding is defective. Automotive electrical shops have equipment to turn commutators and to undercut the mica insulation, although this last chore may be done with a knife blade. Slight abrasions and glaze may be polished out with emery cloth.

With an ohmmeter or test lamp, check the armature for shorts and opens. Connect one probe to the shaft and touch each bar. Resistance between the bars and the shaft should approach infinity. Shorts due to moisture may often be corrected by heating the component in an oven for several hours at about 250 degrees F. Breaks in the windings are repairable if you can reach them with a soldering gun. Fortunately, most opens occur at the point where the commutator bars are soldered to the windings. Extreme overloads heat the generator to the extent that the solder melts and is "thrown." This kind of damage usually means that the armature is beyond repair, but before replacing it, find out what caused the overload. Finally, the armature may be checked with a growler. Auto electric shops have this equipment. A growler is a large transformer in which you place the suspect generator. The growler creates a strong magnetic field. When the suspect armature is rotated manually within the field, the interaction of good segments cutting the transformer field will resist the rotation. At a bad segment, you can feel the decrease in turning resistance.

Perhaps one armature out of a hundred will pass all checks and still refuse to function in the generator. These "flying defects" are caused by the interaction of centrifugal force and magnetism, and give mechanics a certain philosophic cast of mind.

Inspect the field coils for visual damage. Worn spots on the pole shoes and frayed insulation mean that the field screws have become loose or that the shaft bearings are excessively worn. Burnt or carbonated insulation indicates that the field has overheated, and should be replaced.

Using an ohmmeter, check type A coils to see if they have grounded to the frame. By breaking the circuit between them, it is possible to determine resistance in each coil. It is impractical to give the specified reading for all field coils here; usually the figure is between 4 to 8 ohms. Both coils should read within 15 percent of each other. Type B coils are checked for shorts by disconnecting the existing ground.

The generator shaft should have a small amount of end play and no perceptible side play. Normally the drive side

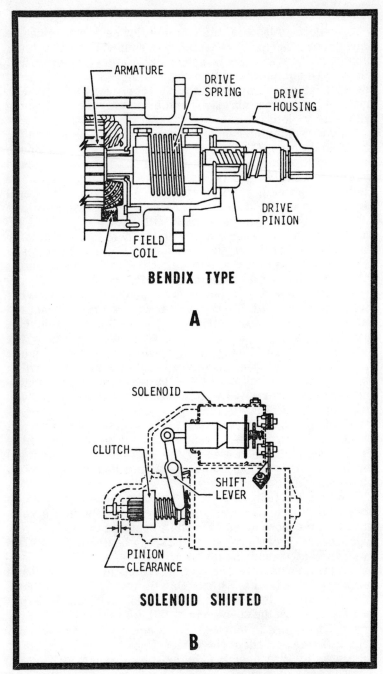

Fig. 4-2. Starter Motor Drives. (Courtesy Kohler of Kohler.)

bearing wears first. Bushings may be driven out with a punch that is slightly larger than the shaft diameter. Often the brush end bushing will be pressed into a blind (closed) boss, and will have to be carefully cut with a chisel. Another method is to ram the bushing out with hydraulic pressure; fill the cavity with heavy grease and position a rod of exactly shaft diameter over the bearing. Strike the rod a sharp blow with a hammer—the grease will drive the bushing out. New bushings are installed by driving them into position with a drift, or better, a vise. No reaming is necessary.

Prior to mounting on the engine, the generator can be made to "motor" by feeding current to it from the battery.
On type A generators:

1. Ground the generator frame to the engine block.

2. Ground the field terminal.

3. Momentarily touch the lead from the ungrounded side of the battery to the armature terminal.
On type B generators:

1. Ground the generator frame to the engine block.

2. Connect the field and the armature terminals with a jumper.

3. Momentarily touch the lead from the ungrounded side of battery to the armature terminal.

STARTER MOTORS

Except for heavier components, the construction of starter motors is quite similar to generators. Wiring is almost identical, and when rotated by an external power source, starters will generate.

Inspection

Starter motors (Fig. 4-2) should be regularly inspected during scheduled lubrication. Look especially for broken springs, and worn teeth. Keep the entire mechanism scrupulously clean. Apply a light coat of oil to prevent rust. The solenoid shifted type may be tested by using a jumper across the heavy (thick) battery terminals.

Servicing

When the engine refuses to crank or cranks too slowly (small engines demand a cranking speed of at least 90 rpm),

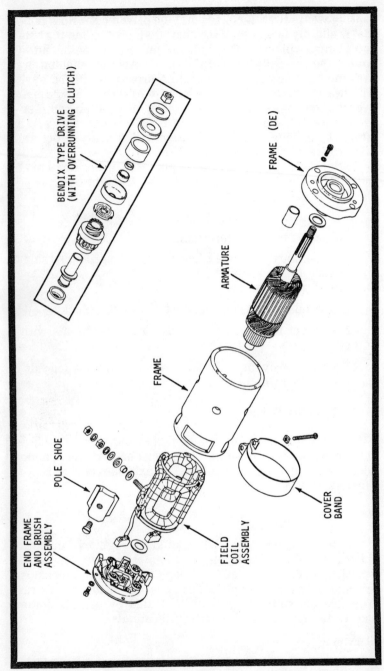

Fig. 4-3. Starter Motor Exploded View. On this type, the brushes can be examined by removing the Cover Band. (Courtesy Kohler of Kohler.)

first determine the charge in the battery. At the battery, the relay, and the starter motor, the cable connections should be scraped bright and tightened. Do not overlook the ground connection. If the cables feel warm to the touch or show signs of mechanical damage, run a resistance test. In no case should the resistance be more than 3 or 4 ohms. If the system includes a relay, jump the heavy battery terminals with a screwdriver. This removes the relay from the circuit. Before dismounting the starter motor from the engine, make one last test. Determine if the flywheel can be turned by hand. Hydraulic lock (oil or water trapped in the chamber, above the piston), rusted piston rings, burned bearings, or jammed starter drive gears can "freeze" an engine.

Once the starter is removed from the engine, inspect the brushes, the drive mechanism, and the bearings (Fig. 4-3). Side play can cause the shaft to bind under load. A bent shaft—not unusual in some designs—can be straightened by means of a pair of vee blocks and a press. The electrical tests for the armature are the same as those for type A generators.

Fig. 4-4 shows a starter solenoid as found on McCulloch and West Bend outboard engines, and on some Japanese motorcycles. It is moisture-proofed.

STARTER-GENERATOR COMBINATIONS

Because starters and generators are mechanically and electrically alike, it is a relatively simple matter to combine

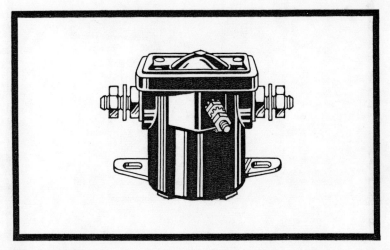

Fig. 4-4. Moistureproofed Starter Solenoid "Black Box." (Courtesy R.E. Phelon Co., Inc.)

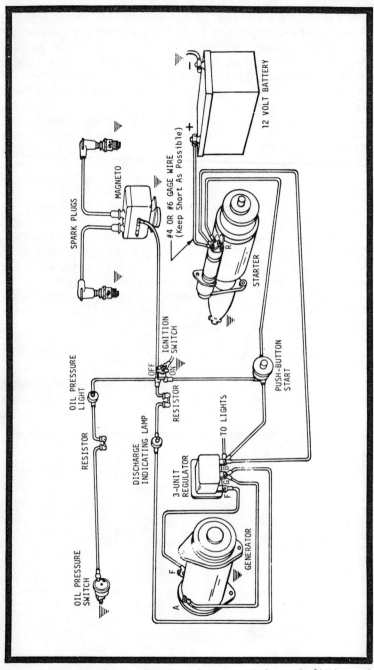

Fig. 4-5. Industrial Engine Electrical System with Separate Starter and Generator. (Courtesy Kohler of Kohler.)

both functions in a single unit. On cranking, current is fed through the brushes, causing the armature to rotate. Once the engine is running, current is then taken off of the brushes, as is done with any DC generator.

One of the earliest starter-generator combinations was the English Siba Dynastart. It had twelve field coils and four brushes (two for the generator function and two for the starter). Some of these units had a rather unique feature—they could be reversed. When the operator of a Villiers three-wheeler wanted reverse, he stopped the engine and started it backwards!

Fig. 4-5 shows a more conventional system as used on American industrial engines. Some of these units employ a solenoid for key operation rather than the mechanical switch shown.

The test procedure outlined here is basically the same as used on Japanese and European designs. Refer to Fig. 4-6. To test the starter:

1. See that the battery is fully charged (a minimum reading of 1.225 on the hydrometer, with no more than .050 variation between cells).

2. Using a good multimeter, switch to the volt function, place the positive probe on No. 2 and the negative probe on No. 1 (ground). The meter should register at least 10 volts. If it doesn't, there is a short in the system.

3. Still holding the negative probe on the ground, move the positive probe to No. 3 terminal. A reading of less than 10 volts means that the cable or the connections are defective.

4. Move the positive probe to No. 4 and depress the starter switch (or energize the solenoid on machines so equipped). Low or no voltage means a defective switch.

5. Move the positive lead to No. 5. Depress the starter switch. A reading of 10 volts or more means that all wiring between the battery and the starter is good. The starter should crank. If it doesn't, the starter is at fault.

To test the generator:

1. Set the multimeter on the 15 amp scale. Disconnect the "B" terminal at the regulator, and connect the meter in series between the terminal and the lead. All current to the battery must pass through the meter. The charge should begin at about 2,000 rpm (sometimes higher on outboards and motorcycles) and the amount of current will vary with the state of charge of the battery. Either of two conditions can occur.

(a) An excessive charge—10 amps or more. Disconnect the lead to the "F" terminal at the regulator. We are dealing with a type A generator and so no current should

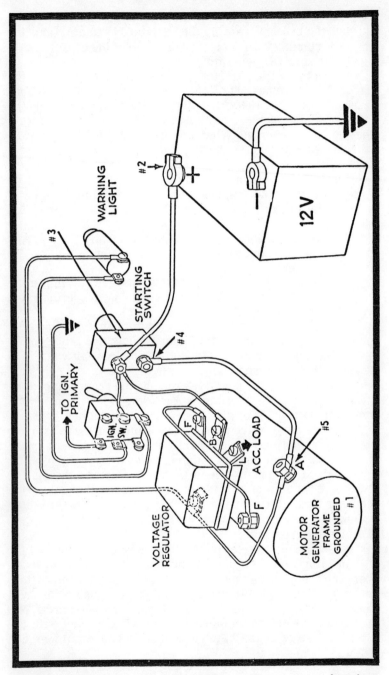

Fig. 4-6. Combination Starter-Generator used on many American Engines. (Courtesy Briggs and Stratton.)

be produced. If the generator still charges, the field coils have grounded to the generator frame.

(b) No or low charge. Short the "F" lead to the ground. The generator should put out 10 amps or more. If it does, the fault is (by elimination) in the voltage regulator. And if it doesn't, the fault must be in the generator, and disassembly and test of individual parts is indicated.

Japanese motorcycles may be equipped with Kokusan or Mitsubishi starter-generators which are driven directly off of the crankshaft and carry the ignition point assembly (Fig. 4-7).

To test the starter:

1. Check the battery condition and see that the terminals are clean and tight.
2. When the key is turned, the solenoid should make an audible click indicating that the field windings are being energized.
3. To check the solenoid contacts, bypass the switch by connecting the battery terminals together.
4. If the starter does not rotate, the problem is in the brushes, field windings, or the armature. Disassemble and test according to instructions under "Starters."

To test the generator:

1. Make the connections shown in the drawing (the dotted lines show how the unit was wired originally).
2. With the field grounded, run the engine at no more than 2,200 rpm. The voltmeter will show 13 volts if the generator is working properly.

To test the regulator:

1. Connect the wiring as shown by the dotted lines.
2. Disconnect the battery from the regulator at "B."
3. Run the engine at 2,500 rpm. A working regulator will show 14.7 to 15.7 volts between its "B" terminal and the ground.

Solenoids

These devices consist of a field coil and a plunger. When the coil is energized, the plunger moves against a spring. Choke controls on outboards are often solenoid operated, and most starter circuits include either a solenoid or its close cousin, a relay. The relay is merely an electro-magnetic switch which may be mounted remotely from the starter. A small current from the ignition switch energizes the relay and allows a large current to flow into the starter windings. When

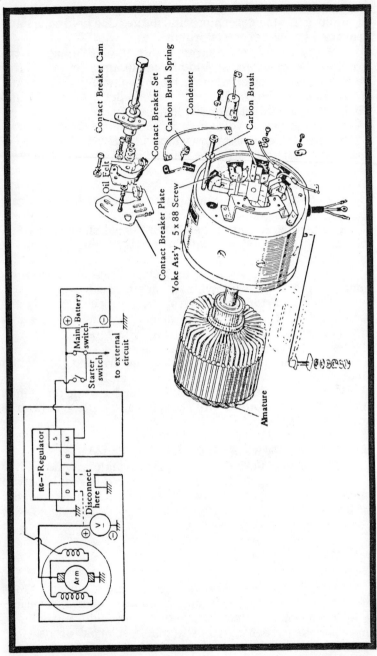

Fig. 4-7. Starter-Generator Combination found on many Japanese Motorcycles. Note that the ignition points operate off of the Armature Shaft. (Courtesy Kawasaki Heavy Industries, Ltd.)

the relay has the additional function of engaging the starter drive as in Fig. 4-2, it is more properly known as a solenoid.

Solenoids and relays should make an audible click when they are energized. The switching function may be tested with an ohmmeter or by jumping the heavy contacts.

ALTERNATING CURRENT SYSTEMS

In the old conception, electric current flows from positive to negative, from the pole where there is a surplus of electrons to the pole where there is a scarcity. In the newer electron theory, electrons flow from negative to positive. In any case, in DC circuits, the polarity remains constant, that is, current flow is unidirectional. In alternating current (AC) circuits, the polarity changes, and the electron flow goes from zero to maximum in one direction, drops to zero, and back to maximum in the other direction. The frequency, stated in Hertz, is the number of cycles per second. Standard American household current has a frequency of 60 Hz and small engine alternators can produce 50,000 Hz.

Alternators

For many years, small engine designers have been moving away from DC systems. Motorcycles are typical in this regard—about the only pure DC systems left are those which employ combination starter-generators. AC generators or, as they are generally known, alternators, are more efficient than their DC counterparts and give better current-voltage regulation.

A magneto is basically a high voltage alternator and it is a simple matter to add low voltage coils on the stator to provide current for lights and other accessories. Fig. 4-8 shows a Briggs and Stratton design which is similar to dozens of others found on industrial engines, motorcycles, and outboards. The large coil provides spark for the ignition and the smaller coils operate the accessories. To check the lighting coils, operate the engine at 2,500 rpm and with all accessories off, and take a voltage reading between the coil output and the ground. On 6-volt systems the reading should be 6 volts or slightly higher.

Rectifiers

AC is not suitable for charging batteries and will quickly burn the point contacts in horns and turn signals. Many small engine electrical systems include a rectifier to convert some

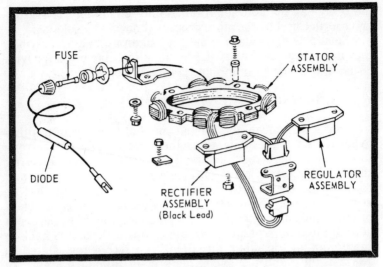

Fig. 4-8. Magneto (High Voltage Alternator) which provides 7 Amps. Note the Rectifier, Regulator, and Diode. The Diode prevents Battery Discharge when Engine is not running. (Courtesy Briggs and Stratton.)

of the AC into DC. A rectifier is an electrical check-valve: current can easily flow through it in one direction, but is blocked by high resistance in the other. The selenium rectifier (Fig. 4-9) consists of a steel disc (or square) coated with selenium alloy. Current will flow from the steel to the alloy, but not in the opposite direction. Silicon rectifiers, sometimes called diodes, are akin to transistors and are cylindrical in shape. This type of rectifier is gradually replacing the selenium disc.

In the wiring diagram of Fig. 4-9, note the ballast resistor (top sketch) in the ignition circuit. The ballast limits current flow to the coil. In the starter solenoid circuit (third from the top), the limit switch prevents the starter solenoid from operating when the power transmission is engaged. In other words, the limit switch here is doing electrically what a so-called Bendix spring would do in an automobile starter.

A single rectifier in series with an AC source will block half of the current. In order to rectify all of the current coming off of the alternator, a bridge circuit consisting of four rectifiers is required. This is known as full-wave rectification and is used on many motorcycles and outboards. The gain over half-wave rectification is obvious. Full-wave selenium units are stacked on a central bolt with terminals connected to each plate. Silicon rectifiers are usually encapsulated in a "black box" which may have cooling fins.

While there are many timesaving techniques for testing rectifiers on different engines, the surest test is to reproduce the current passing—current blocking function. Most full-wave selenium and silicon units have three terminals. The center is DC to the battery and the outer two are AC from the alternator. Using an ohmmeter, set on the lowest (1½ volt) range, hold one probe on the center tap and the other on an AC input. The rectifier should either pass current or block it. Reverse the leads and the opposite condition should exist. Repeat the test with the other AC input.

Be very careful when working with rectifiers, especially diodes. Vibration, reversed battery polarity, or too much test voltage, will instantly ruin them. When soldering diodes, always keep the leads as long as possible and use a heat dam to protect the component. A pair of long-nosed pliers clamped to the lead with rubber bands helps to block the heat. If the unit is faulty, it may be replaced with a factory part or with an equivalent.

VOLTAGE AND CURRENT REGULATION

DC generators are normally fitted with an external voltage and current control. An unregulated generator will put out voltage and current roughly in proportion to engine speed, and will quickly destroy itself and other components in the system. Alternators have a kind of built-in current limiting factor, but, with the exception of a few industrial engines, require some means of voltage control.

Mechanical Regulators

The earliest type of regulator was entirely mechanical, and consisted of a vibrating reed switch. Current or voltage depended upon the relative amount of time the switch remained open.

When servicing this type of regulator, be certain that it is well-grounded to the engine and that the contact points are clean. Use a fine-cut riffle file on the contacts and finish with carbon tetrachloride or TV tuner cleaner (in a well ventilated working area). Never use emery cloth or sandpaper. All mechanical regulators have provision for adjustment, either by means of a screw or by bending a tab. To check the voltage output, run the engine for a few minutes to stabilize the system and disconnect the battery at the "B" terminal and the ground. Run the engine up until the needle stops flickering. This is the cut-out voltage and should be between 7.5 to 8 volts

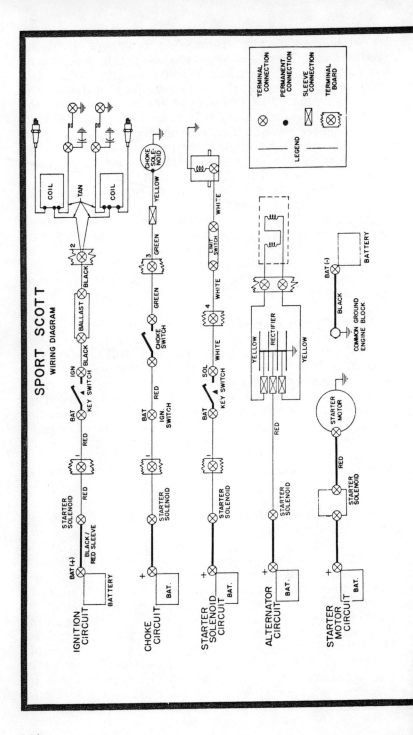

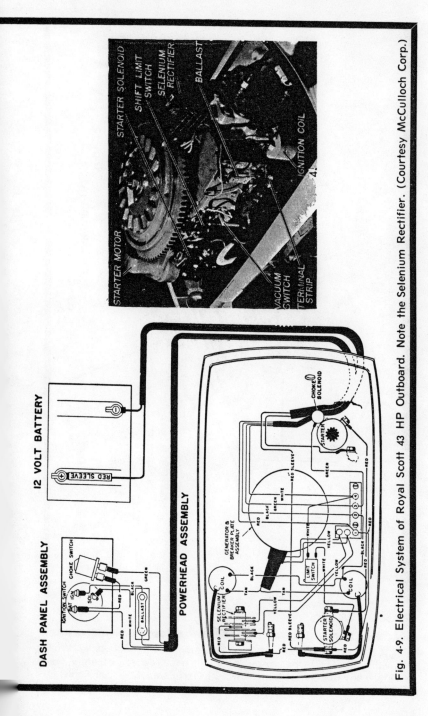

Fig. 4-9. Electrical System of Royal Scott 43 HP Outboard. Note the Selenium Rectifier. (Courtesy McCulloch Corp.)

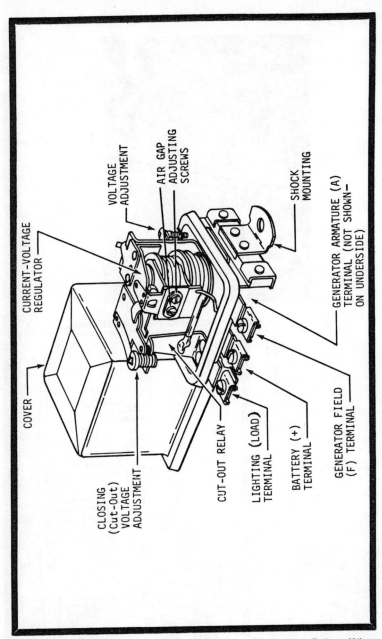

Fig. 4-10. Typical Current-Voltage Regulator with Cut-out Relay. When adjusting voltage output, make sure unit has reached operating temperature and disconnect battery from + terminal. (Courtesy Kohler of Kohler.)

on a six volt system and between 14 and 15 volts on a twelve.

Direct-current generators must be isolated from the battery with a cut-out relay. Otherwise, the battery would discharge through the generator when the engine was not running. Alternators do not require a cut-out since the rectifier performs the same function. Points in relays are subject to sticking and in time, can get out of adjustment. Set the relay so that it closes at one half volt over rated battery voltage.

There are many misconceptions about voltage regulators, the chief one being that once you replace a defective regulator the job is done. More often than not, a defective regulator is only a symptom of some other electrical trouble. Unless you find the trouble, the new regulator will also fail.

Examine the defective unit. Burnt points or discolored springs on the cut-out unit mean that the generator was not polarized. Burnt points on the voltage and current units can mean high resistance in the charging circuit, or a bad ground. Burnt windings indicate extremely high resistance or an open circuit. If the damage is limited to the current points, suspect a short in the system.

When installing a new regulator, be certain that it is one specified by the manufacturer. Each regulator is carefully tailored to its generator-alternator and a substitution will guarantee further electrical problems. One point of difference is load resistors (built into the bottom of the regulator), another is polarity, and another is in the fundamental circuitry of Type "A" and "B" designs. The regulator should have a perfect ground against bare metal and should be mounted away from engine heat. If the unit chatters on starting, immediately stop the engine. The generator polarity is wrong.

A typical current-voltage regulator is shown in Fig. 4-10.

Electronic Regulation

The best mechanical regulators have a relatively short life. The ideal solution is a solid-state regulator with no moving parts. On some motorcycles, a single Zener diode controls the amount of current and voltage delivered to the battery. A Zener may be likened to an electrical pop-off valve: one terminal is connected to the alternator, the other to the battery, and the base is grounded to the frame. As battery voltage climbs (which should indicate the approximate state of charge of the battery) the Zener diverts excess current to the ground. At approximately 15 volts on the battery, all current goes to ground.

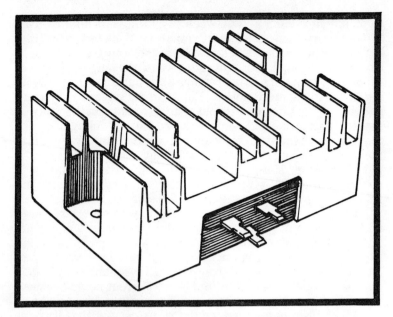

Fig. 4-11. Regulator-Rectifier "Black Box" used on John Deere and Onan Engines. (Courtesy R.E. Phelon Co., Inc.)

To test a Zener, connect a voltmeter across the battery terminals and an ammeter in series between the alternator and the Zener. Operate the engine at 2,500 rpm and observe the meters. At 13.5 to 15.3 volts, the Zener should be conducting 1 amp to the battery. As battery voltage increases, this current should drop to zero.

Some of the more expensive American and Japanese engines use transistorized regulators which may be combined with a rectifier in a single "black box." Fig. 4-8 shows a Briggs and Stratton unit with a separate regulator. Fig. 4-11 is a drawing of a typical Repco "black box" which combines both functions. It is simple enough to test the Briggs and Stratton unit, but test procedures for the Repco type of combination unit can become quite involved. The problem is compounded since current and voltage values differ with the application. If you have problems with one of these units, the best advice is to obtain a factory manual for the particular engine. Many mechanics—even dealer mechanics—"test" these "black boxes" by means of substitution. But the following general procedures may be helpful:

1. On three-terminal designs, the center tap is almost always DC to the battery.

2. Connect an ammeter in series with the battery terminal—the meter will register the charge going to the battery and to whatever accessories are on the circuit.

3. With the engine running at 2,500 rpm, observe the amount of charge, and once it has stabilized, turn on the lights and other accessories. (Those units without accessories may be loaded with a G.E. no. 4001 headlamp.) The rate of charge should increase. If it doesn't, the problem is in the regulator—rectifier or in the alternator.

4. To test the alternator, connect the lamp across the two AC leads. It should burn brightly at medium engine speeds.

FUSES AND CIRCUITBREAKERS

Most electrical systems contain a fuse or a circuitbreaker on the main line to protect against excessive current draw. Chronic difficulties here mean either a short in the system, a defective circuitbreaker, or that fuses with a current rating below the manufacturer's specification have been installed.

LEAD-ACID BATTERIES

On small engines, batteries have two functions. The first is to provide energy for the starter motor and the second function, which is often overlooked, is to provide voltage regulation. The engine should not be operated for extended periods with the battery out of the circuit.

Each cell consists of a number of positive and negative plates, held in position by means of glass fiber separators (Fig. 4-12). A glass mat is sandwiched between the separator and the positive plate in order to hold the active material (lead peroxide) in place. The case is usually made of high-impact plastic and, in the smaller sizes, is transparent, so that the plates can be inspected. The case is filled with a mixture of sulphuric acid and water. Each cell produces approximately 2.1 volts. Thus a 6 volt battery will have three cells, a 12 volt battery will have six.

The capacity of the battery is a function of the plate area and is rated in ampere-hours. A 10 A-H battery should discharge 1 amp continuously for 10 hours. But note that this figure does not mean that the battery will produce 10 amps in 1 hour! There is a limit to the amount of current which can be taken from a battery at one time. This is why batteries are rated on either a 20 hour or a 10 hour scale. When replacing a

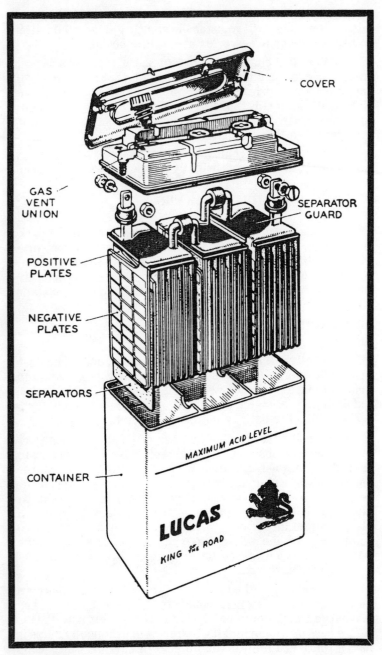

Fig. 4-12. Construction of Lead-Acid Battery used on British Motorcycles. (Courtesy Royal Enfield.)

battery, be certain that the new battery has at least the same A-H rating on the same time scale.

Servicing Batteries

Batteries require careful, periodic maintenance. See that the case is clean, because deposits on the case increase the rate of self-discharge. Periodically remove, clean, and re-tighten the terminals (including the ground to the engine and starter connections). After the battery terminals have been tightened, add a coat of grease on the outside of the terminal to help prevent corrosion. Add distilled water as needed. If one cell is consistently low, it is leaking and the battery should be replaced.

A typical battery can be expected to lose 1 percent of its charge per day. If the charge falls to 75 percent of its rating, the battery may be permanently damaged. Therefore, do not allow a battery to stand idle for more than one month.

Small batteries should be put on a trickle charger for 24 hours or so. Service station "hot shots" are just about guaranteed to destroy small batteries through over-heating of the electrolyte. As a battery is being charged it gives off considerable quantities of hydrogen gas which is highly explosive. Charge batteries in a well-ventilated area, observe the no smoking rule, and **never** connect or disconnect hot cables. That is, turn the charger on after the cables are connected, and off before they are disconnected.

The surest way to tell if the battery has an adequate charge is with the hydrometer. A fully charged battery will give a reading of 1.270; a dead battery will give approximately 1.040.

Battery Chargers

While not an integral part of small engine repair, a faulty battery charger can certainly give a mechanic headaches. Briggs and Stratton has devised a simple tester which is shown in Fig. 4-13. Parts needed to build the tester are:

1. One each 1N5061 diode.

2. Two lamp sockets, such as Delco No. 0931-102 (red) and No. 0932-102 (green).

3. Two No. 53 bulbs (for twelve volts).

4. Two ¾" screws to serve as terminals.

If the charger is working properly, the green bulb will light. If both bulbs light, the rectifier diode in the charger is defective,

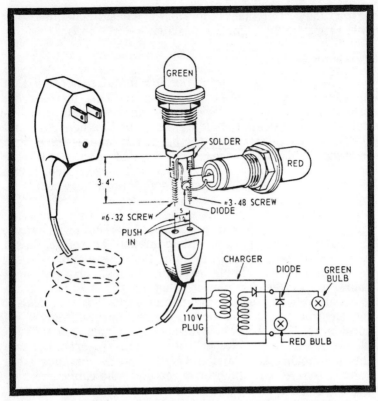

Fig. 4-13. Battery Charger Tester. (Courtesy Briggs and Stratton.)

and if neither lights, the line cord, switch, or transformer is defective.

WIRING

Except for spark plug leads and battery cables, the wiring on small engines is known as primary wire. The conductor is copper, which may be tinned for easier soldering and better conductivity. Nearly all primary wire consists of 7, 19, and in the larger sizes, 37 individual strands inside of plastic insulator. Stranded wire is preferred because it is flexible and resists vibration damage. About the only use for single-conductor wire is in certain magnetos where the wire is very close to moving parts. A single-conductor wire is stiff and will hold its shape and location.

Most primary wire is vinyl insulated and is rated at 600 volts. Vinyl is good for 105 degrees Celsius before it melts.

Certain high temperature applications might require teflon insulation which tolerates 200 degrees C. and costs about twice as much as vinyl. The old-fashioned rubber and cotton insulation is still available from Belden for people who restore vintage automobiles, but it has no advantage for modern small engines.

Selecting Wire

A conductor is selected by its resistance. Excessive resistance will cause overheating and voltage drop. Three factors are involved: the cross-sectional area of the wire, the length of the wire, and the amount of electrical load. The cross-sectional area conforms to a standard known as the American wire gauge (AWG). Each diameter is assigned a number so that the cross-sectional area is approximately doubled every three numbers. The lower the number, the larger the diameter of the wire. Thus No. 10 wire has twice the area of No. 13. And because resistance is inversely proportional to cross section, No. 10 will have only half the resistance of No. 13. Thus, for a given current load, heat loss and voltage drop will be cut in half for No. 10 as compared with No. 13 AWG wire.

The length of the wire is also important. You can appreciate this if you have ever tried to rig a hundred-foot drop cord. There will be almost no power coming out of the end of the wire. Most of the electrical energy has been used to overcome resistance along the length of the wire. The easiest way to solve the problem is to go to a larger diameter conductor.

And finally there is the factor of load, which is expressed in watts. Volts times amperes equals watts (EI equals W). A lamp which draws 4 amps at 6 volts is rated at 24 watts. But if the voltage is doubled, the current draw can be cut in half for the same amount of wattage. Amps—the quantity of electrons going past a given point—are what determines the resistance. If we cut the amps in half, the cross-sectional area of the conductor can also be halved without increasing the resistance. This is why 12-volt systems have smaller diameter wires than 6-volt systems.

When rewiring, you should use at least the same diameter as the original. But sometimes new wiring will be added to the system, as when installing remote controls on outboards. Here is a chart which you can use to select the proper gauge. To use the chart, measure the length of the wire to the load (on two-wire systems, double the length), select the proper amperage

Amperage		Candlepower		Wire Gauge and Length					
6 volts	12 volts	6 volts	12 volts	2'	5'	10'	15'	20'	25'	30'
.05	1.0	3	6	18	18	18	18	18	18	18
1.00	2.0	5	10	18	18	18	18	18	18	18
1.50	3.0	12	24	18	18	18	18	18	18	18
2.00	4.0	15	30	18	18	18	18	18	18	18
2.50	5.0	20	40	18	18	18	18	18	18	18
3.00	6.0	25	50	18	18	18	18	18	16	16
3.50	7.0	30	60	18	18	18	18	16	16	16
4.00	8.0	35	70	18	18	18	18	16	16	16
5.00	10.0	40	80	18	18	18	16	16	16	14
5.50	11.0	45	90	18	18	18	16	16	14	14
6.00	12.0	50	100	18	18	18	16	16	14	14
7.50	15.0	60	120	18	18	18	14	14	12	12
9.0	18.0	70	140	18	18	16	14	14	12	12
10.0	20.0	80	160	18	18	16	14	12	10	10
11.0	22.0	90	180	18	18	16	12	12	10	10
12.0	24.0	100	200	18	18	16	12	12	10	10

or candlepower, and move under the footage column to find the correct wire gauge.

Rewiring

Before you begin a major rewiring job, make certain that you understand the function of each component. It helps to make a drawing first, even if you have the factory schematic. Often the original wiring can be simplified and you can add convenience and safety features such as multiple fuse boxes. However, a word of caution here: if the machine is un-fused, as are some Harley-Davidson motorcycles, a fuse might cause trouble. Usually the factory engineers know what they are doing, and if they don't protect a circuit, it is because the circuit surges in normal operation and would blow a fuse. In these cases, a piece of fusible wire inserted on the "hot" battery cable might be a better solution. This wire is available from GM dealers. Another point to consider is that heavy current draws should be taken from junctions which can supply the current. The starter solenoid or the "B" terminal on the voltage regulator are good sources of current.

As a general rule, the routing of the wires should be as direct as possible both to reduce the resistance and the amount of wire used. However, the wires ought to be tucked out of the way to avoid damage and should never be so short that they are under tension. Of course, you should route around exhaust manifolds and other hot spots. Install grommets wherever the wires pass through bulkheads or firewalls. Bundles of wire can be taped together or, better, can be installed with shrinkable tubing. This tubing will shrink as much as 50 percent when heated. It is available from radio supply stores.

All connections should be soldered for good conductivity. The "shady tree" practice of twisting wire together will work for a time, but eventually will result in a high resistance joint. Use 50-50 rosin core solder and a 100-150 watt soldering gun. The conductor should be heated so that the solder flows into the joint. Overheating will cause the solder to lump and is as bad as no solder at all. In the past few years, solderless connections have become popular. They are convenient and are reliable. But the individual cost is higher and a special crimping tool is required. Splicing can be done with a connector or by soldering the wires together. The approved method is to strip the insulation back about ¾" on both wires and splay the strands. The strands are then interleaved and twisted, and the splice is soldered. All joints should be taped (most mechanics prefer plastic tape). And, in marine applications, the tape should be coated with silicone cement to prevent moisture from entering the joint.

Various styles of solder and solderless terminals are available. Use the closed-end type for permanent connections and the open-end for ease of removal. Typically, the leads to the voltage regulator will be open-ended and those to the coil primary will be closed-ended.

Chapter 5
Engine Service

Before undertaking any major mechanical work, make a thorough inspection. The condition of the spark plug(s) can tell a great deal about what is going on in the engine (see Chapter 2) and can help to organize the work by allowing you to anticipate problems. An excessively oily plug means that the cylinder in question needs ring or valve work, and one that shows marks of overheating means that you may expect internal damage. A compression test is also useful as a diagnostic aid and as a reference point to check your work after assembly. On multi-cylinder units, all cylinders should be within five or six pounds of each other. If pressure in one cylinder is excessively high, expect carbon build-up from faulty oil control in that cylinder. If compression for one or more cylinders is low, then the problem is rings, valves, or possibly, the head gasket. An ounce or so of oil poured into the cylinder should markedly increase compression if the rings are bad. If the oil has no effect, the problem is a compression leak in the upper cylinder area, and a head removal may be all the teardown that is necessary.

On four-cycles, the dip-stick can tell a story. Thick, carbonized oil almost invariably means that the rod bearings have burned. Hold the wet end of the stick up to the sun: any metallic particles in the oil will sparkle in the light. On new and just rebuilt engines, the presence of metal in the oil may not be critical; but on older engines, it means a complete teardown is necessary to discover which parts are wearing. Water can accumulate in the crankcase from normal condensation (especially in marine engines) or from leaks in the cooling system. In heavy concentrations, moisture will turn the oil greyish white. A good test for water is to touch the dip stick to a hot exhaust manifold. Water will separate out and burst into steam. Expect bearing damage if the condition has been chronic, or if the water contains ethylene glycol antifreeze. A strong odor of gasoline in the oil normally indicates blowby around worn rings and pistons. However, remember that gasoline can also enter the crankcase through a ruptured fuel pump diaphragm.

If you can move the flywheel from side to side, across the axis of the crank, the main bearings are worn and should be replaced. This is not an overly difficult repair to make on industrial and motorcycle power plants. But to replace the mains on lawnmower engines requires considerable labor and a fairly heavy investment in reamers and boring jigs. Most shops simply tell the customer that he needs a new engine.

Play in the connecting rod bearings shows itself in free movement of the crankshaft. Turn the engine to the compression stroke and rock the flywheel a few degrees either way in the plane of rotation. If the crank can be moved without moving the piston, the conn rod bearings are bad. Sometimes it might be necessary to remove the head in order to observe the piston directly.

Major work should be done in a clean, well-lighted area. All parts, including the fasteners, should be cleaned in kerosene or Varsol as they are removed. Have separate containers for each major assembly so that you will not spend time later wondering which part goes where. Superficially identical parts, such as pistons and rocker arms, should be tagged to prevent confusion during reasembly.

The sections that follow describe engine components in a top to bottom sequence, with appropriate fault isolation and repair descriptions.

CYLINDER HEADS

The primary function of the head is to trap the energy of combustion so that it will be exerted on the piston. Obviously, the head must be physically strong enough to resist distortion. The big problem is the interface between the head and the block. Most engines use a gasket which may be made of soft metal such as aluminum or copper or, more commonly, of some asbestos compound. The placing of the bolts is also important—on engines which continually blow head gaskets, the difficulty is with the location of the fasteners.

High-performance engines sometimes do not have a head gasket. The head and the cylinder surfaces may be carefully machined and spigoted (as on Porsche industrial engines), or both head and cylinder may be cast in one piece (Power Products' two-cycles are a good example of this method.) Without a gasket, reliability is increased and most importantly, heat transfer is enhanced. The head, which forms the roof of the combustion chamber, is the most critical hot spot in the engine. Heat must be allowed to move away from it, either downward into the block, or, better, out of the engine entirely via cooling fins or water passages. The Suzuki water-cooled 750 incorporates both fins and water passages in its head. Better cooling is also why engineers have come to employ aluminum heads on small engines.

The shape of the combustion chamber is also very critical, especially in overhead valve designs. The engine will only run as well as it "breathes." The more air-fuel mix which can be "inhaled" on the intake stroke, the more power will be developed at the flywheel. And the easier the exhaust gases can escape, the less energy will be lost in pumping (back-pressures). On two-cycles, the gas flow must change direction several times as it enters the crankcase and passes up through the transfer ports and out the exhaust. But a well-designed four-cycle can have an almost straight-line gas path from the carburetor venturi to the muffler. Currently, the most sophisticated designs feature a hemispherical chamber, with the flat top of the piston forming the floor, and with the intake and exhaust valves at a large angle to each other. Most Honda cylinder heads are hemispherical.

Removing the Cylinder Head

Once you have determined that the head must be removed, proceed as follows. First clean the engine with steam or with a mix of concentrated Gunk and solvent. The Gunk can be washed off with high pressure water. Now remove all the parts that attach to the head, such as the generator bracket, the intake and exhaust manifolds, throttle assemblies, etc. On air-cooled engines, the head bolts are often masked by shrouding. Working from the center outward (to prevent distortion) remove the bolts. Note that some bolts may be longer than others; mark these to avoid confusion at reassembly. On overhead valve designs, you will usually have to remove the rocker arms and the push rods to get at some of the bolts. On overhead cam engines, use a center punch to mark the position of the cam relative to some fixed reference point. It is also a good idea to mark the position of the crankshaft as well, since someone might disturb the engine before the head is assembled. On some ohc engines you can save yourself a great deal of trouble if you secure the driving chain by a piece of wire or twine. Otherwise the chain will have to be fished out of its case.

Once the bolts are loose, the head should slip off. If it doesn't, disconnect the ignition and crank the engine. Compression should pop the head off. Sometimes this tactic fails on liquid-cooled engines with aluminum heads. The combination of water, aluminum, and cast iron causes electrolysis which binds the head to the block. The ultimate tactic is to install the head bolts finger tight and start the engine. **Never** insert a pry bar between the head and the block or use a steel hammer to shock the head loose.

The head should be lifted straight up, so that the gasket surfaces are not scratched. If a gasket is fitted, it should not be used again. The fiber types become compressed and lose their ability to conform to irregular surfaces, and the solid metallic types harden with age and heat. In an emergency, however, you can re-use a head gasket if it is not fractured. Brush liberal amounts of Kopper-Kote on both sides of the gasket (being careful to keep the excess out of the chamber). Copper and aluminum gaskets can be heated with a propane torch and quenched, so that the metal will become soft and ductile.

Inspection and Analysis

Inspect the head bolts for pulled threads and for signs of bottoming. Metal slivers on the threads may mean that the boss is stripped. Some mechanics take the precaution of running a tap down each head bolt boss in order to reform the threads and to clean them for accurate torque readings during assembly.

The interface between the block and head ought to be dead parallel. In practice, some distortion is tolerated since the gasket can fill in any slight irregularities. To check for distortion, obtain a machinists' straightedge and lay it across the gasket face of the head. If you can insert a .004" feeler gauge at any point under the straightedge, the head is beyond tolerance for industrial engines. On motorcycles, the maximum allowable distortion is .002".

Warped cast iron heads require milling, although small heads can be turned on a lathe with a four-jaw chuck. Small aluminum heads can be ground with a homemade rig. Tape a piece of wet-or-dry sandpaper (about No. 240 grade) to a flat surface. A drill press work table is ideal, but you can also use a piece of plate glass. Hold the head at a central point, and work it back and forth across the abrasive until the gasket surfaces are uniformly shiny. Keeping the paper wetted with oil will increase the rate of cut. Finish with paper of about No. 340 grade. However, you should never take more than .010" off a head. To go beyond this means that the compression ratio has been significantly increased beyond what the maker intended. And although there will be some power increase, an engine modified in this way may require premium fuel, revamped ignition timing, and cooler spark plugs.

The kind and the amount of carbon in the chamber can point out problems in other parts of the engine. Heavy carbon deposits means excessive oil is being burned. On two-cycles, this condition may come about by an improper fuel-oil mix (either from the operator's carelessness or from evaporation of the gasoline in storage) or an overly rich carburetor adjustment. But a word of caution: before you decide that a two-cycle has accumulated too much carbon, check it against a similar engine. Some low-speed two-cycles need to be decarbonized every 50 operating hours. On four-cycles, excessive carbon in one or more cylinders means that oil is entering the chamber. Probably it is coming from below, passing by worn or stuck rings. Or it can enter from above on overhead valve designs, due to worn valve guides and leaking valve seals. The problem will be intensified if the breather is stopped up, or, on dry-sump motorcycles, if the oil pump is not timed correctly.

Black, sooty carbon is a sign that chamber temperatures have been low, and is not a cause for alarm. Heavy yellow deposits suggest that the brand of gasoline should perhaps be changed. Chalk white means that the engine has been run very hard, or that it has overheated.

Carbon deposits may be removed with a wire brush (Fig. 5-1) or with a dull knife. Be careful not to damage the chamber and the gasket surfaces. Scratches will cause leakage and make carbon removal more difficult the next time. Mechanics who are interested in extracting maximum performance will sometimes polish the intake passages with a Dremel grinder. The object is to speed the gas flow by removing casting marks and by smoothing curves. On multi-cylinder power plants, the chamber volume is equalized by grinding and welding.

Finally, inspect the head for cracks. Cracked fins are not a serious problem, but a cracked water jacket can be disastrous. Often these cracks will be internal and it may be necessary to have the head pressure tested at the local automobile machine shop. If there is evidence of heavy corrosion, remove the welch (freeze) plugs which are on most water cooled designs, and have the head boiled out. These plugs are themselves a source of trouble. They are designed to "blow out" if the coolant freezes, thus protecting the jacket. They can develop leaks and they tend to rust from the inside out, where it is not obvious. Always replace them with new welch plugs.

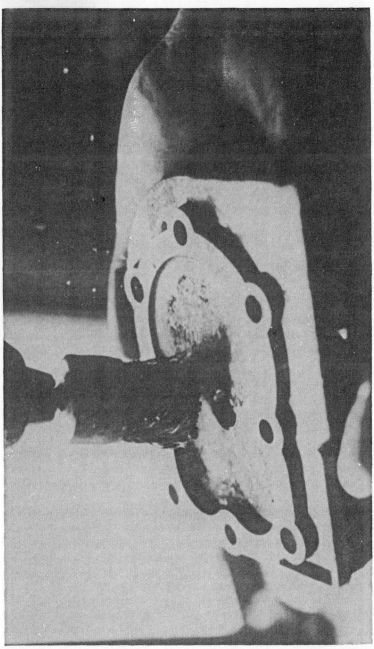

Fig. 5-1. Removing Carbon with a Wire Brush. You can also use a screwdriver or a knife, but be careful not to scratch the gasket surfaces. (Photo by Al Galinski.)

Replacing the Cylinder Head

Set the gasket in position, noting whether it has a top or bottom. Normally no gasket compound is used, as this will only make the engine more difficult to disassemble the next time. Install the head and run the bolts up finger tight. Then, with a torque wrench, tighten the bolts to factory specifications. More accurate readings are possible if the fasteners are lubricated with high pressure grease before assembly. The amount of torque varies with the diameter of the bolt or stud, with the material into which it is screwed, and with the manufacturer. Most lawnmower engine head fasteners are designed to be tightened to 140 inch-pounds, and the small (up to 90cc displacement) Hondas average 100 inch-pounds. The larger Kohler K331, which develops 9½ HP, requires 480 inch-pounds on the head. It is impossible in a book of this length to list all torque figures for all makes and models which you might encounter. The best advice is to call the distributor for this information as the need arises.

In addition there is a definite torque sequence to be followed on each head. This is to prevent the head from warping as it is brought down. The general principle is to work from the center fastener out to the ends of the head. Thus, the sequence would be: bolts near the center; bolt to the right of center; bolt to left of center; 2nd bolt to the right of center; 2nd to the left, etc. As you approach the torque limit, work in increments of five to seven pounds, so that the head will be stressed more or less equally. After the first hour of operation, the head should be allowed to cool, and then re-torqued.

VALVES

Other than the ignition and carburetion ills which plague all engines, the most troublesome feature of four-cycles is the valves. There have been many experiments with disc or barrel valves, but few of these designs have been commercially successful. The English Crossley industrial engine built around the turn of the century was perhaps the best of the lot, but it has long since disappeared. Today all four-cycles use the familiar tulip-shaped poppet valve.

The head of the valve carries a milled surface on its outer edge called the face. Older engines may have the face cut at 30 degrees to the stem, but experiments with flow benches have shown that the contemporary 45 or 46 degree angle is more efficient. When the valve is closed, the face makes a gas-tight seal against a ring of hard metal called the seat. On cast iron

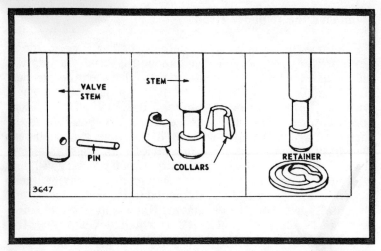

Fig. 5-2. Three Types of Valve Spring Keepers. (Courtesy of Briggs and Stratton.)

side valve engines, such as the Clinton and the Briggs, the seat may be nothing more than a chamfered hole in the block. Cast iron is durable enough to make a reasonably satisfactory valve seat, at least for low-speed engines. But on high-speed engines, the seat is made of a special heat- and corrosion-resistant steel (CRES). It is usually shrunk into place. (Honda motorcycles are unique in that they have a steel skull which covers the whole combustion chamber and which is relieved to form the valve seats.) The width of the contact area is determined by the machined surface on the seat. Ideally, the contact area should be knife-sharp, but this is impossible in practice. A wide seat would be good from the point of view of heat dissipation, but is liable to carbon fouling. Each manufacturer has his own suggested seat width, which varies from .040" on the 50cc engines to no more than 1-16" on larger engines.

Heat is transferred from the valve head to the seat when the valve is closed, and through the stem to the valve guides when it is open. The intake valve has an easier time of it, since the incoming charge helps cool it. Intake valves are usually made of chrome nickel alloy steel. The exhaust valve may have a Stellite or Silchrome face or be constructed entirely of XCR or stainless steel. The exhaust valve is often of smaller diameter than the intake because exhaust gases exit under relatively high pressure, while the intake side in un-supercharged engines is limited to one atmosphere (14.7 pounds per square inch).

Removing Valves

After the head is removed on side valve engines, the crankshaft should be rotated until the piston is near top dead center on the compression stroke. Both valves will be closed. To remove the valves, it is necessary to compress the spring. You can do this by jiggling two screwdrivers, or you may use a valve spring compressor designed specifically for small engines. The Briggs and Stratton part number 19063 is one of the best. For very small engines, the tool will have to be modified by grinding the ends slightly so that it will fit into the valve chamber. On ohv designs, the springs can be compressed by the screwdriver method or by a C-clamp tool such as Harley-Davidson 65-36. Once the spring is compressed, the keepers can be removed (see Fig. 5-2). The valve should lift easily out of the guide. If it sticks, polish off the accumulated carbon and varnish on the stem with strip emery cloth. Clean the valve thoroughly with a wire wheel and insert it back into the guide. Wobble means that the guide has worn. A loose guide will make it difficult for the valve to seat and will cause high oil consumption on ohv and ohc engines.

Replacing Valve Guides

Most small engines have replaceable bronze guides. A few slide valve designs run the valve directly on the cast iron block. On an engine like this there are two options open: either replace the valve with one that has a slightly larger stem or else fit a bronze guide (Fig. 5-3). In either case, special reamers are required. To install new guides on an engine which has bronze guides, drive the old part out with a drift or pull it out with a screw extractor. The new guide then can be driven into position with a soft mallet. Some mechanics speed the process by first chilling the guide with dry ice. Finish reaming will be required, however, since it is almost impossible to install a guide without doing some damage to the bore. Suitable reamers can be purchased from the factory or, in inch sizes, from large hardware stores. Expansion reamers seem like a good buy, because one reamer will handle several sizes, but are generally frowned upon by careful mechanics because they do not have the accuracy of the fixed type. When using any reamer, always turn clockwise, in the direction of the bite of the teeth. Back-turning will quickly dull the tool.

Checking Valve Wear

The valve should be chucked up in an electric drill and checked for straightness. It is difficult to get a gas-tight seal

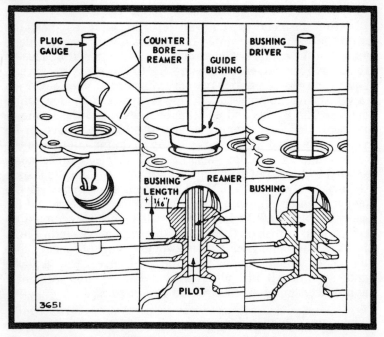

Fig. 5-3. One Procedure for Installing Valve Guides. On other engines, the valve guide can be pressed out, and a new one (chilled with dry ice) dropped into place. (Courtesy Briggs and Stratton.)

with a valve that is even slightly bent. Inspect the valve face and the seat. Minor irregularities can be cleaned up by lapping.

Lapping Valves

Years ago, valves came with slots or a pair of holes in the head. A tool was inserted and the valve was positively rotated. Today valve crowns are machined smooth to prevent "hot spots," and all that can be done is to turn the valve with a suction cup. This works well enough on the larger engines, but no one has yet made a really good suction cup tool in the smaller diameters. K-D part number 501 is as good as any, but in my experience at least, will more or less continually keep coming unstuck. One partial cure is to use a little silicone cement on the outer edges of the cup. Once the valve is secure, the lapping is quite simple. Compound comes in small double-lidded tins. One side is coarse for the initial work, the other half is fine grade for finishing. Dab a small amount of compound on the face of the valve and rotate the suction cup with the palms of both hands in the manner of a Boy Scout making a fire. Periodically replenish the compound and turn the valve so that the whole circumference is lapped. When finished, you should be able to hear a dull "thump" as the valve is dropped on the seat. Admittedly, this takes a good ear. The Navy suggests that the valve face be coated with Prussian Blue (available at the larger hardware stores), then the valve is let down on its seat, and rotated one quarter turn with a slight downward pressure. If the lapping has been done correctly, a uniform and unbroken ring of blue will appear on the seat. Perhaps the surest method is to dab a small amount of liquid soap around the valve. Hold the valve down with finger pressure and apply compressed air to the back of the valve by way of the port. Bubbles mean leakage.

Valve lash or clearance is set to specifications, either by adjustment screws, or, on most side valve engines, by grinding the stem. With the cam in the down (valve closed) position, insert a flat feeler gauge between the end of the stem and the cam follower (Fig. 5-4).

Heavily pitted faces and seats must be machine ground to tolerance. Any reputable automobile machine shop can handle this work for a nominal sum. Have the machinist check the seat angle before he begins work, since the angle can vary with different makes and models. It should not be necessary to lap the valves afterwards.

Valve Seats

If loose, the seat may be peened as shown in Fig. 5-5. But if the seat is deeply pitted or cracked, it must be replaced. Purchase or make a puller like the one pictured in Fig. 5-6 and remove the old seat. New seats are driven into place with a special driver and pilot, although in a pinch you can use an old valve. As already mentioned, some cast iron engines do not have seats as such. The valve rides directly on the block. To install a seat on these it is necessary to use a factory supplied reamer and pilot.

Valve Springs

Except for Ducati motorcycles, all small four-cycle engines employ springs to close the valves. The Ducati design (called "desmodromic") was developed in their racing machines several years ago and is now a feature of their touring models. The valve is at all times controlled by the cam, both an opening on closing. With this arrangement, there

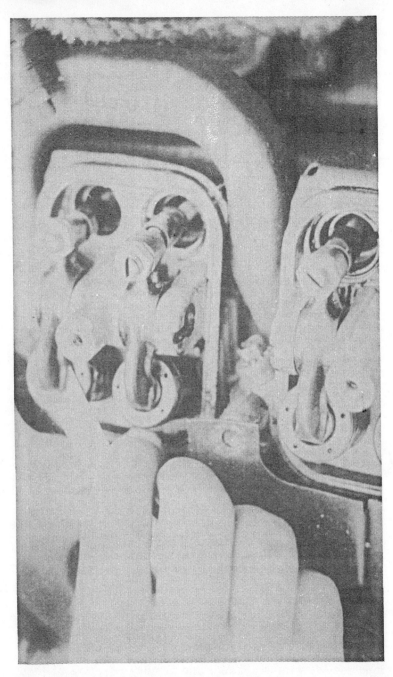

Fig. 5-4. Checking the Valve Clearance on a Two-cylinder Continental Engine. Note the rocker arms and pushrods. (Photo by Al Galinski.)

is no valve float, and consequent loss of power at high rpm is prevented.

Valve springs are made of corrosion-resistant steel and come in various styles. Some air-cooled engines have used "clothes' pin" springs which twist rather than compress. Since the coil is remote from the valve stem, these springs are somewhat insulated from heat. But if the spring breaks, the engine is out of commission. A variation of this design is the torsion bar spring used on some imported motorcycles. One end of the bar is anchored to the head, and the free end is fitted with an arm which moves with the valve. As the valve opens, the bar is twisted.

The most common type of spring is a simple coil which works in compression. These springs are relatively inexpensive to manufacture and the technology is very well understood. But at high rpms, strange things happen. The center coils, which are farthest from the supports at either end, may compress more than the rest of the spring. Moving out from the center, a surge wave of compressed coils travels up and down the length of the spring at about the speed of sound. At

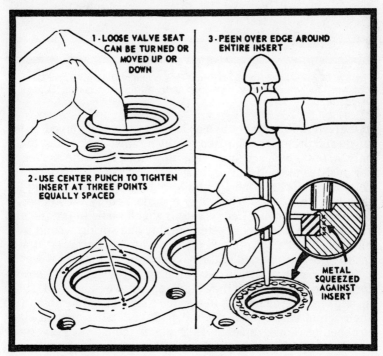

Fig. 5-5. The Procedure for Peening a Loose Valve Seat. (Courtesy Briggs and Stratton.)

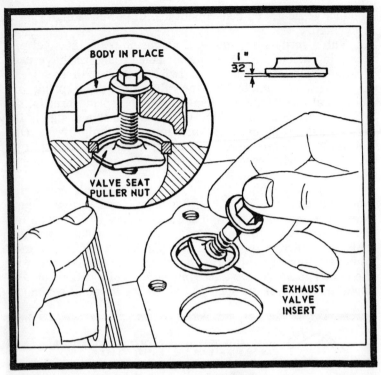

Fig. 5-6. A Valve Seat Removal Tool. This is one of the many special tools that may be fabricated in the shop with minimal expense. (Courtesy Briggs and Stratton.)

any given moment only two or three coils are involved in the surge and these are collapsed solidly against each other. Since only a few coils are involved, the tension of the spring is seriously weakened. Surge also causes the spring to break: the collapsed coils have inertia and want to keep moving along the length of the spring. When they are blocked by the keeper, a terrific hammering effect is set up which leads to fatigue and rapid failure. Engineers often employ one spring wound over another to reduce surge and try to design the natural, "tuning fork," vibration period of the spring to be as high as possible.

The intake and exhaust valve springs usually do not interchange, even though they may appear to be identical. If the springs have grown weak in service they will not be able to do their job of closing the valve at high rpm when inertia loads are heavy. Some mechanics replace the springs at every overhaul as a form of insurance. However, springs can be reused if they meet the manufacturer's specifications. Two measurements are involved: the spring height and the spring tension. Height is determined by measuring the free (uncompressed) length of the spring against specifications. To measure the tension requires a special tool, often sold as an attachment for a torque wrench. The spring is inserted in the tool and compressed until the coils are solid. The tension, in inch-pounds, is then compared to the reading given by a new spring. It is possible to increase the tension of a spring marginally by installing it with stacked keepers, although this practice is not recommended.

Inspect the spring for wear between the coils. Wear here can only mean that surge has occurred, and the spring should be replaced. Pitting is also cause for rejection.

PUSH RODS AND ROCKER ARMS

Overhead valve engines employ push rods to open the valves. Frequently, the inside of the push rods serve as the oil passages for lubrication of the rocker arms. It is good practice to tag each rod as it is removed so that it will be replaced in its own valve assembly. Like old shoes, engine parts get comfortable with use. And on a few engines the intake and exhaust valve push rods are different lengths. Incorrect assembly can cause valve-piston collision. Roll the pushrod on a flat surface to see that it is straight, since even a slightly bent rod will rob power by not opening the valves fully. Inspect the ends of the rod for galling and for discoloration from overheating. If the rod is hollow, blow it out with compressed air.

The rocker arms on small engines are steel forgings which ride on bushings or needle bearings. Rocker arms wear on three places—the adjusting screw, the point where the valve touches the arm, and the bearing. If the screw is worn or has mushroomed, it is generally better to replace it than to attempt to regrind. These screws are only surface hardened. The point of valve contact can be reground with the proper equipment—do not attempt to do it by hand. A correct regrind will result in near silent operation. A worn bushing can be replaced either from dealer stock or through a bearing supply house. Be sure that the oil hole is aligned properly. After it is pressed into place, ream to a clearance of .001".

Valve Lifters

Sometimes called cam followers or tappets, valve lifters convert the rotary motion of the cam into reciprocating motion. There are two basic types: mechanical and hydraulic. The latter is coming into wider use in small engines since it

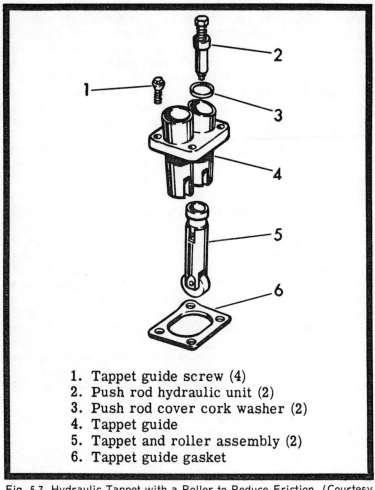

1. Tappet guide screw (4)
2. Push rod hydraulic unit (2)
3. Push rod cover cork washer (2)
4. Tappet guide
5. Tappet and roller assembly (2)
6. Tappet guide gasket

Fig. 5-7. Hydraulic Tappet with a Roller to Reduce Friction. (Courtesy Harley-Davidson.)

has several advantages over the mechanical type. With a well-functioning hydraulic lifter, there is no lash in the valve actuating mechanism, and consequently, none of the "tap-tap" noise associated with ohv engines. The lifter automatically compensates for variations in lash as the engine heats up and parts expand, and, once the lifter is adjusted, it can be ignored for the life of the engine. However, a hydraulic lifter is a precision device that can be damaged by careless handling and by dirty oil.

The hydraulic lifter consists of a cylinder and a piston which is sometimes known as the plunger (Fig. 5-7). These two parts are individually fitted at the factory to extreme tolerances, closer than any other engine part. The bottom of the cylinder has an oil port drilled in it. This port is controlled by a ball check valve and a spring. When the lifter is on the heel of the cam, the ball check valve moves off its seat because of oil pressure supplied by the pump. The cylinder fills with oil. As the lifter rides upward on the opening flank of the cam, the check valve closes and oil is trapped in the cylinder. The lifter then opens the valve because the column of oil between the cylinder and the piston is not compressible. Hence, the name hydraulic lifter. But if the oil remained trapped in the lifter, the valve would stay open. This is where the controlled leak-down rate comes into play. A metered amount of oil is allowed to escape around the piston so that the valve can seat, but not so much that there will be any lash in the mechanism.

Lifter Noise

The first sign of lifter trouble is a tapping noise which persists after the engine has run for a few minutes. Some starting noise is normal with a cold engine because the lifters have bled down. But the noise should stop as soon as the lifter has had a chance to charge. To check, remove a lifter from the bore and disassemble. Do not take apart more than one at a time since the parts are not interchangeable. The piston is held in place by a spring clip. Turn the lifter on its end and allow the piston to fall out. If it is stuck in the bore because of varnish, you can tap the face of the cylinder on a block of hard wood. Below the piston is the ball check valve and spring. Be careful with these parts, as on some designs they are loose and may get lost. Clean the parts thoroughly in solvent and inspect for scratches on the piston and the cylinder bore. Scratches or wear marks will increase the leak-down rate and cause noisy operation. Inspect the ball with a magnifying glass. In most cases, a thorough cleaning is all that is required.

To test the lifter, place it in a container of clean 10W oil and work the plunger with a screwdriver. Five or six strokes should fill the cylinder. Now apply heavy pressure in the range of 60 to 75 pounds on the screwdriver attempting to drive the piston down. It should hold this pressure for at least 15 seconds without moving more than $1/8$ of an inch. More movement than this would indicate that the leak-down rate is too high and that the lifter should be replaced. Some large automobile repair shops have leak-down testers.

Before installing, blow out the oil ports which feed the lifter, and then charge the device as above. Failure to do so might lead to piston-valve interference because of air trapped in the lifter. Most lifters are preset at zero lash. Turn the adjustment screw with the valve seated until you can no longer turn the push rods with your fingers. On Harley-Davidson twins, the screw is lengthened four turns after zero lash.

Mechanical lifters are solid slugs which may be mushroom shaped (as in all American industrial engines) or which may have a roller to reduce scuffing, as in Fig. 5-7. The mushroom types are generally slightly off center to the cam, so that the valve tends to rotate as it is opened. Mechanical lifters are the last part to wear out in an engine, although the face where it contacts the cam should be inspected for scoring.

PISTONS

Years ago, nearly all small engine pistons were cast iron. These pistons gave exceptional wear characteristics, especially if they were tin coated, and ran very quietly since they could be mated to the bore with close tolerances. However, the demand for higher rpm has made the cast iron piston a curiousity. Today, all small engine pistons are made of cast or forged aluminum.

Aluminum has about a 20 percent weight advantage over cast iron. This weight saving translates into less energy loss in overcoming the inertia of the piston, and in lighter loads on the crankpin bearings. Aluminum is a much better conductor of heat than is cast iron, and makes possible lower combustion chamber temperatures. However, as it is heated, aluminum will expand rapidly. This growth must be allowed for in the design of the piston. Most aluminum pistons are cam ground, that is, they are oval shaped in cross section. When the engine is cold, only the ends of the oval touch the bore; as it reaches operating temperature, the piston fills out and contacts the bore on all points. Some pistons have T-slots in the skirt which give space for the metal to expand without seizing. And, in general, aluminum pistons are set up looser than the cast iron equivalents. Consequently, most of these pistons "slap" when cold.

Structurally, the piston consists of a head or crown which forms the floor of the combustion chamber. The crown is relatively thick to provide the needed strength (combustion pressures approach 1,000 pounds per square inch), and to provide for heat dissipation. The rings are fitted in grooves below the skirt and above the wrist pin. The wrist pin is fitted to a heavily reenforced boss which may carry replaceable bronze bushings. In the larger engines, the pin is a press-fit in the boss and does not move relative to the piston. All the movement is confined to the small end of the connecting rod. Smaller engines have full-floating wrist pins; the pin moves on the conn rod, and is free to move on the piston bosses as well. Snap rings or teflon pads locate the pin at either end of the boss, so that it cannot move laterally and contact the cylinder bore. The lower part of the piston, called the skirt, acts as a bearing to keep the piston and the rings parallel to the bore.

Removing Pistons

On motorcycles and outboards, the piston can be removed by lifting the cylinder barrels and taking out the wrist pin. On industrial engines, the connecting rod is disconnected at the big end, by removing the sump or the side plate. The piston and the upper portion of the rod are pushed upward out of the bore as one assembly.

Remove the spring clips (or "circlips" as the British term them) with long-nosed or spring-clip pliers. Be careful not to drop the clips into the crankcase. If the piston is removed by disconnecting the conn rod, the rings must slide over the cylinder ridge. This ridge will be about a quarter of an inch below the top of the cylinder, and is caused by wear. The bottom of the ridge represents the limit of travel of the top ring. When removing the piston from a worn engine, the rings will often hang on this ridge. Forcing the piston will only damage it. Push the piston back down into the bore and remove the ridge with a reamer. Automotive parts houses sell these tools, which are adjustable to a wide range of cylinder sizes. The better ones have carbide cutting edges. When reaming a cylinder, it is a good precaution to seal the edge of the piston and the bore with masking tape. The shavings then can be blown out with compressed air.

If the wrist pin is stubborn, it may be removed with heat and a soft drift pin. I said that small engines have free floating pins, but they float only when the piston is at operating temperature. There are several methods used to apply heat to the piston. The sloppy way is to wrap the part with rags soaked in boiling water. Most mechanics prefer to use an electric hot plate. A propane torch may be used if the flame is kept moving over large areas of the piston. The danger in this method is local overheating and consequent distortion. Drive the wrist pin out far enough to clear the small end of the conn rod. Check it for wear and signs of overheating. The pin rarely gives trouble in modern engines, but it is a heavily stressed part and should be viewed with respect.

Most four-cycle pistons have a front and back, as indicated by the letter "F" or an arrow stamped on the crown. On the older two-cycles with domed pistons, the steep side of the dome must face the transfer port. If assembled wrong, the incoming charge will tend to go out the exhaust port. And when handling pistons, remember that they are fragile. Dropping one on the skirt usually ruins it.

Severe overheating will crystallize the aluminum crown, giving it a dull, grainy appearance. Pistons with this kind of damage must be replaced and the cause of the overheating should be found. Usually it will be an ignition timing error, an air leak, the wrong spark plug heat range, or a too lean carburetor adjustment. The underside of the piston should be inspected for cracks, particularly around the wrist pin bosses. Any carbon on the underside of a two-cycle piston should be removed. Scratches on the skirt mean that the oil has not been changed regularly on four-cycles, or that the crankcase breather is stuck open, allowing dust to enter the engine. On two-cycles, scratches mean dirty (usually rusty) fuel, air leaks in the crankcase or in the induction tract, or else that the filter has failed. As a general rule, dust can enter wherever liquid leaks.

Determining the Problems

Piston wear is best determined with the aid of inside and outside micrometers. The bore is measured at several points and the diameter of the piston is subtracted, leaving the clearance. Experienced mechanics can estimate the wear by the amount of wobble present near the top of the bore. On high-speed engines, the maximum clearance is .004". Low-speed engines are less demanding and can go to .005". More clearance than this means that it is hardly worthwhile to install new rings since the "fix" will only last a few hours. If the bore is only slightly worn and is still round, it is possible to take up some slack by installing a new standard sized piston. Sometimes it is practical to knurl the old piston, or, on the larger engines, to install a steel expander in the skirt. But the safest bet is to bore the cylinder to the next oversize. These oversizes are available in .010", .020", and .030" for American engines, with a few manufacturers offering up to .060" oversize. The oversize will be stamped on the crown of the piston as 10, 20, 30, etc.

Occasionally you will encounter an early model engine which has since been up-dated by the factory with a larger bore. Assuming that the same castings are used, you can bore the engine to accept the new, larger pistons. In this case, the pistons might be .080" or .120" over the stock original. However, a word of caution: with this modification, torque will go up dramatically, but so will the compression ratio. An engine which was happy on regular might require high test fuel, retarded ignition, richer main jets, etc. In other words, increasing cubic inches can get you into trouble.

Rebuilding Pistons

After determining the condition of the piston, clean the ring grooves. There are tools available for this chore, but a broken ring mounted in a file holder is adequate. Be careful because the edges are razor sharp. If you are cleaning a number of pistons and have the time, it helps to soak the ring grooves with paint remover for a few days. Sears' brand seems to work the best. After all traces of carbon are removed, take a new ring and fit it to the top groove. This is the groove which will wear first, since it is closest to the combustion chamber. More than .003" side play on high-speed engines and more than .005" on the low-speed, industrial types means that the ring will flutter and break. Normally the piston must be replaced, although Perfect Circle makes spacers which can be installed on the larger pistons. Some lathe work will be required to prepare the groove to accept the spacer.

New pistons are normally purchased from the manufacturer who built the engine. However, it may be possible to substitute other brands of piston in certain applications. For example, a mechanic may wish to use a forged piston for its greater strength and durability, rather than the cast stock item. The critical parameters are the bore size, the diameter of the wrist pin, and the height of the crown above the pin. The last measurement determines the compression ratio. There should always be at least .120" between the crown of the piston and the valves (open) on ohv engines. This clearance allows for carbon build up and for high speed stretch of the conn rod and valve train. It may be necessary to cut valve clearance notches in the crown of the piston. Flycutting is preferable, although expensive if you do not have access to a milling machine. A cheaper, but slower, method is to fix carbide cutting edges to a discarded valve. The cutters should be arranged radially around the face of the valve so that they cut a circle equal to the valve diameter. They are secured with epoxy cement or with silver solder. The piston is installed and the head torqued down with the valve loose in its guide. Then with the stem of the valve chucked in a drill motor, take a series of light cuts. The head will have to be removed to check the work and for each notch.

RINGS

Piston rings have three major functions: they must seal compression above the piston (and below on two-cycles), transfer heat from the piston crown to the cylinder walls, and, on four-cycles, must deliver a metered amount of oil to the cylinder. All of these functions must be done with a minimum of friction, and with some ability to adjust to out-of-round and tapered bores.

Sealing

Sealing is the result of two factors—the metal to metal contact of the face of the ring with the bore and the pressure applied to the ring. Any metal surface, even one that has been polished with new techniques such as Micro-Finishing, has irregularities. When two metal surfaces are in contact, these irregularities interlock and produce heat with friction. The rougher the metal, the more friction will be produced. But after an extended period of contact, the irregularities will smooth out with a kind of mutual honing action. When this has taken place, we say that the parts have been "broken in" or seated. If the two metals are of the same hardness, the seating-in period will be extended. But if one is relatively soft, it will smooth out rather quickly. Therefore, the ideal piston ring should be made of metal that is either softer or harder than the bore. Most engines have bores of cast iron, which is not really a hard metal in terms of the choices available today. For rapid seating and best durability, piston rings ought to be constructed of some extremely hard metal such as nickel (stainless) steel. These rings force the cylinder to conform and may have double the life of the more conventional rings.

But stainless and equivalent alloys are quite expensive and difficult to work. The market for them is generally limited to premium engines which involve heavy labor charges during overhaul. Some marine and racing engines use these rings.

Manufacturers of small industrial and motorcycle engines generally specify cast iron. The material is cheap and requires no heavy investment in foundry and machine tools. On the other hand, cast iron is the material used for most cylinder bores, and this means that the rings will be slow in seating and will give relatively poor wear characteristics. However, during the war, engineers learned to apply a flash coating of chrome to aircraft rings. Chromium is an extremely hard corrosion-resistant material and is an ideal wearing surface for rings. Many small engines now use chrome plated rings. Harley-Davidson chromed all their rings until they discovered that the oil rings outlasted every friction surface in the engine.

A more recent technique is to apply molybdenum (or "moly") to the upper rings. This material has good wear characteristics and, paradoxically, seats quickly. It is sprayed into a groove around the face of the ring (hence the term "moly-filled" rings) and is bonded with heat.

But no surface seal is adequate by itself. There must be a positive force pushing the ring against the cylinder wall in order to achieve a leak-down rate of 10 percent, which is considered the absolute maximum. A well-set up engine might have a leak-down of only 3 percent.

All conventional rings are spring-loaded in the sense that they would expand to a larger circle than the cylinder diameter if not held within the bore. This is why a compressor tool is needed to install rings. The force pushing the ring open may only be a few ounces, but it is critical. When it is lost from fatigue, breaks in the ring, or from accumulations of varnish which hold the ring tight in the groove, engine performance suffers.

The compression rings have an additional boost in the form of gases which escape around the sides of the piston and exert pressure behind the ring. Some compression rings are wedge-shaped profile (as on Evinrude outboards) to encourage gas flow, while others have steps as shown in several of the illustrations in Fig. 5-8. Some new designs such as the head-land and the Dykes ring depend almost entirely upon compression for sealing.

Ring sealing can also be improved by using extremely narrow rings. This gives high pressure (force per area), but relatively little friction since the contact area is small. Because of wear considerations, these rings are always chrome plated.

Oil rings are an additional complication on four-cycle engines. The ring closest to the bottom of the piston picks up oil and distributes it along the bore. The lower drawings in Fig. 5-8 show several styles of one piece oil rings. Sometimes another oil control ring is fitted above the first. This is the scraper ring which has the form of an "L" turned on its side. It keeps excessive oil from flooding the bore and finding its way into the combustion chamber. Some European engines, which are designed for long life between overhauls, may have an

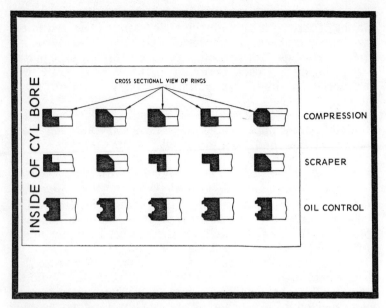

Fig. 5-8. Cross-Sectional Views of the Compression, Scraper, and Oil-Control Rings. It is extremely important that the rings be installed as shown. (Courtesy Briggs and Stratton.)

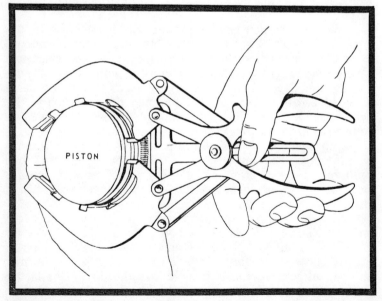

Fig. 5-9. Using a Piston Ring Expander. This is the safe and easy way to fit rings, although your fingers may be used if you are careful not to twist the rings. (Courtesy Briggs and Stratton.)

additional ring at the base of the piston, below the oil ring. This guide ring looks like a compression ring, but actually only functions to keep the piston parallel in the bore.

A piston ring expander is convenient to use when removing and installing rings, but you can use your fingers (Fig. 5-9). The edges are very sharp, so be careful. New rings should be checked for clearance in the piston groove (see above) and for end gap. Tamp the ring down square into the bore with the flat end of the piston. The ring should rest near the bottom of the bore where the diameter is little affected by wear. Measure the gap between ends of the ring with a feeler gauge. It should be no less than .002" per inch of bore diameter (Fig. 5-10). For example a 3" bore requires an end gap of .006". Excessive gap encourages blowby and insufficient gap can cause the rings to break in service. There must be room for the rings to expand when they get hot. If the gap is too narrow, gently file the ends.

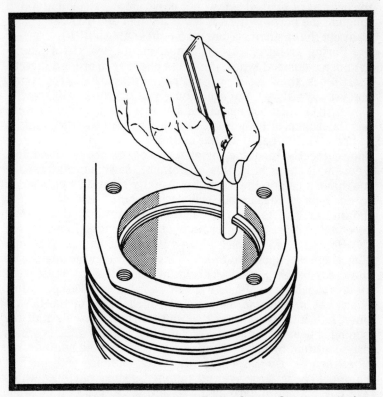

Fig. 5-10. Measuring End-Gap with a Feeler Gauge. On worn cylinders, this measurement should be taken near the bottom of the bore. (Courtesy Kohler of Kohler.)

Rings come in standard sizes and .010", .020", and .030" (American) oversizes. A few companies make an .050" oversize for worn cylinders, but never go any higher without boring. Replacement ring sets may be available in two styles—"factory" or "engineered." Factory rings are the same as originally specified when the engine was new. Engineered rings are designed to have high side pressures, which keeps the rings in contact with out-of-round, or tapered bores. Because of the additional friction, fuel economy might drop 5 percent or so.

Cylinder Grinding

Before new rings are fitted, the cylinder must be ground lightly with a hone. This is to break the glaze so that the new rings will seat rapidly. If honing is omitted, the engine will continue to smoke and burn oil until the rings finally seat. Use the hone as described below, but only remove enough metal to roughen the bore—a hone is a cutting tool and you can easily wear out the customer's newly re-built engine with it.

Boring is fairly rare in small engine work. On single-cylinder industrial types, it is often cheaper to order a factory short-block (new block, piston, rings, crankshaft, rod, cam, and usually valves). But occasionally, a customer will come in who insists upon boring, either because the cylinder is worn past tolerance, or to increase torque. The best way to do this is to use a boring bar with the cylinder mounted in a lathe. If done correctly, machine boring does not require finish honing. Few small shops have this equipment, however. Automobile machinists can do the work, but usually cannot be persuaded to do it for a reasonable price, since small engine work is unfamiliar and takes a great deal of "set-up" time. Most mechanics must use an ordinary spring-loaded hone.

Honing, even simple "glaze-busting," is an art. The final finish depends upon two factors: the grit size of the stones and the cross hatch pattern. The latter should be at 30 degrees which translates to about 70 strokes a minute at 400 rpm. Most shops use an electric drill motor which, even in the heavy-duty, half-inch size, turns double this speed. As a result the honing is poor. Cutting stones come in a variety of grits which are identified by series numbers. The higher the number, the finer the grit. Do not confuse the series number with the grit number. The latter is average number of stone particles which will cover a one square inch average. A 280 grit cutting stone consists of particles which are 1-280" in area. This is equivalent to a series 500 stone.

As a rule of thumb, the harder the ring material, the finer the finish cut should be. Cast iron rings set best with a 200 series stone, while chrome requires a 300, and stainless a 500. The series number identifies the finish. Thus when mechanics talk of a "300" finish they mean that a 300 series stone has been used.

Another aspect of honing is the cutting oil. All petroleum-based oils will dissolve the adhesive which binds the stone particles together. Cutting action is reduced and the stones will have to be frequently replaced. If you use lubricant at all, use animal fat, vegetable oils (such as Crisco), or commercial honing oils. Some mechanics prefer to hone without any lubricant at all. This gives rapid cutting action and only a slightly rougher finish. But once a stone has been oiled, it must continue to be lubricated.

Fig. 5-11 shows the set up for boring. Do not tighten the jig to the table—some movement will help keep the hone cen-

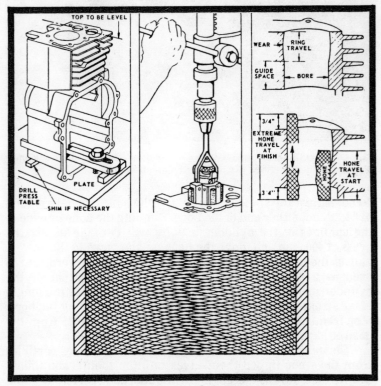

Fig. 5-11. The Set-Up used to Re-Size a Cylinder, and the Desired Cross-Hatch Pattern inside a Honed Cylinder. Scrub the cylinder with detergent and hot water to remove all the stone particles. (Courtesy Briggs and Stratton.)

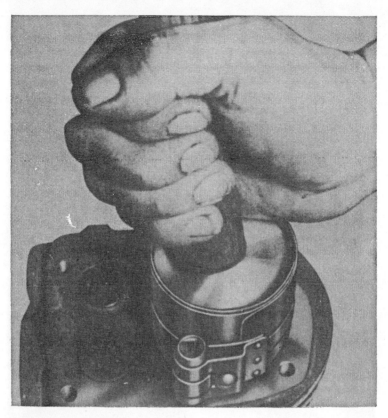

Fig. 5-12. Installing a Piston with the aid of Ring Compressor in a Kohler Single-Cylinder Engine. (Courtesy Kohler of Kohler.)

tered. Adjust the spindle lock so that the hone can only extend ¾" to 1" out either end of the bore. Running the hone outside of the confines of the cylinder usually will damage it. Set the press at 400 rpm and move the hone up and down in the bottom half of the cylinder. This part of the bore is not worn, but is used as a guide for the upper part. As the bottom of the cylinder increases in diameter, gradually increase the length of the strokes, until the hone travels the full length of the bore. Stop frequently to check the progress. At about .0015 from the desired diameter, change to a light finish stone.

Cleaning the bore after honing is critical. Tiny particles from the stone imbed themselves into the metal, and unless they are removed, the engine will undergo a period of rapid wear. The common practice of swabbing the bore with a rag soaked in solvent is not adequate. Solvent actually floats the particles deeper into the metal. Instead use hot water—as hot as you can stand it—and detergent. Scrub the bore thoroughly with a stiff brush and wipe dry with paper towels. Repeat the process until the towels go through the bore without discoloration. This means that all the abrasive particles have been removed. Finally, coat the bore with oil to prevent rust.

A ring compressor is needed to install the piston. Two models are available. The standard model is for engines with one-piece crankcases and cylinder bores (Fig. 5-12). The piston on these engines goes in from the top. The second type of compressor is used on engines with separate cylinder jugs. To assemble these engines, the jug must be brought down over the piston and the compressor must be made to come apart so that it can be removed. These compressors can be purchased from outboard motor parts distribution centers, while the first type can be had from Briggs and Stratton (part No. 18070) or from some of the larger automobile parts houses.

Either type is used the same way. First the crown and ring area of the piston is liberally coated with oil. Then the compressor is tightened around the rings to squeeze them into their grooves. A hammer handle or a rawhide mallet is used to drive the piston (or jugs) home. Studs which protrude out the end of the rod should be covered with lengths of fuel hose to prevent metal to metal contact with the crankpin. The piston should progress into the bore smoothly. If it sticks, find the problem (which might be a ring insufficiently compressed, or interference between the rod and the crank shaft).

CYLINDERS

Ideally the cylinder should be a perfect circle with zero taper. It should be constructed of some relatively hard material such as steel. Cast iron is acceptable if the castings have been aged prior to final machining. Stresses set up in the foundry operations must have time to relieve themselves before the cylinder is trued. Because of the critical nature of the cylinder bore and because of the tolerances necessary for long engine life, some manufacturers of high-performance engines hold cylinder tolerances to .0001", but lawnmower engines and the like are built to plus or minus .001". Cylinder liners are sometimes used in cast iron engines. Of course, a liner or chrome plating is necessary to provide a wearing surface in aluminum engines. These liners, or sleeves as they are sometimes known, are usually cast iron which has been cooled as it is spun at high speed. Centrifugal force helps give a uniform grain characteristic to the metal.

Liners and Sleeves

On liquid-cooled engines, the liner may be wet or dry. Dry liners are shrunk fit into the water jacket. No water is in direct contact with a liner. Wet liners, which are used widely on small European engines such as the Fiat and Renault, are surrounded by water. The liners are positioned by the head and are sealed at the lower end by a composition gasket and copper ring. The gasket keeps water out of the crankcase. To change liners on these engines it is only necessary to remove the head and lift the liner out of the bore. Engine overhaul is considerably simplified.

Air-cooled engines may also employ sleeves. The American practice is to leave the outer circumference of the liner as cast, and pour the cylinder metal around it. Removal is not practical in small engines, nor are replacement liners available. The European practice is to shrink fit the liner on aluminum engines. These liners can be replaced, although the work is a bit tricky. Most shops refuse to do it; and generally, dealers prefer to sell a new cylinder which may run as much as $125, compared with a liner which rarely costs more than $15. But a mechanic should learn how to do this job, if only for his own satisfaction.

First, obtain the liner. You may have to write to the regional or national parts distributor to get one. On the more obscure brands, you might have to write to the factory and pay air freight. But the liners are available.

When the part arrives, inspect it carefully for shipping damage. These liners are never more than a quarter of an inch thick and are quite fragile, as I learned when I dropped one. Check the outside surface for burrs and lathe tool marks which might make installation difficult.

The old liner can be removed because aluminum expands more quickly when heated than does cast iron. Place the cylinder barrel in the kitchen oven and support it by bricks so that it does not rest on the metal floor of the oven. Set the oven at 500 degrees F. and let the cylinder cook for several hours. After you are satisfied that it is uniformly heated, remove the cylinder with tongs or leather welding gloves. The liner should fall out. On most engines, the liner is removed (and installed) from the top. It may be necessary to rap the cylinder sharply against a hard surface to shock the liner loose. As long as you are in the kitchen, you can use your wife's cutting board. But if the liner is warped as from severe overheating or from a broken rod, it can be driven out with a brass drift. Be careful not to scratch or gouge the aluminum. Return the cylinder to the oven, to keep it expanded and ready for the next step.

Sleeve Replacement

The replacement sleeve should be cooled down to the lowest temperature available. A mixture of dry ice and alcohol does the best job, but you can store it in the home freezer for several hours. It may be helpful to use a non-petroleum based lubricant on its outer surface. White lead is most readily available.

The cold liner and the hot cylinder should fall together. The only problem is that you will have to work quickly, since once the parts touch there will be a tremendous amount of heat transfer. At the outside you will have ten seconds to locate the liner to the exact depth of the old one. On two-cycles, there is the additional complication of ports: these must be in dead alignment. But if you fail, and the liner sticks half way down the bore, all is not lost. Simply repeat the process.

And when you're all through, scrub the stove and kitchen, dress up, and treat your wife to a good meal at a restaurant!

For a number of years there has been concerted effort by manufacturers to find an alternate solution to the wear problem in aluminum engines. A liner may add as much as 2 pounds of dead weight to each cylinder. This dead weight is merely a bearing surface and contributes nothing in the way of power. The interface between the liner and the aluminum cylinder acts as a heat dam and makes cooling more difficult. Two alternatives have been used.

Briggs and Stratton has developed a series of small engines which have chromed cylinders. The chrome is plated directly on the aluminum bore and is hard enough to provide a wearing surface for the rings. The chrome is not like the mirror surfaces on decorative items, but has a dull, uneven surface. These surface irregularities form pockets which hold the oil. As added protection from scuffing, the piston is also chromed. Various English modifications of the Villiers "Starmaker" motorcycle engine also feature chromed bores. The only disadvantages are that the chrome may peel (in which case the cylinder must be discarded), and, of course, the cylinder may not be bored oversize.

Another approach is exemplified by the Chevolet Vega automobile engine. The block is die-cast aluminum and the cast iron piston rings ride directly on it. There are no sleeves, nor is the bore chromed. The "trick" is that the block is made of aluminum with a very high silicon content. Silicon is one of

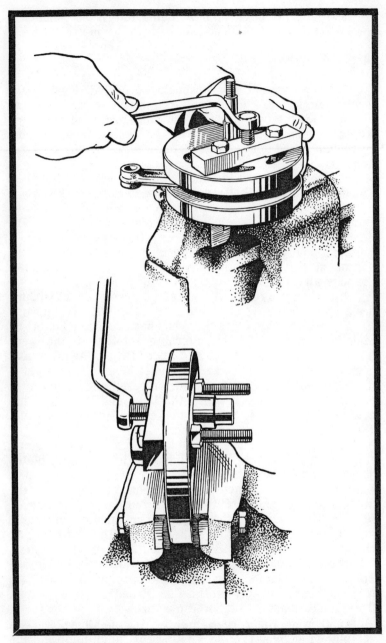

Fig. 5-13. Separating the Flywheels and Removing the Crankpin on a Triumph Tiger Cub. Assembly is the reverse procedure, but the wheels must be aligned with a dial indicator or with special factory gauges. This is no job for a beginner. (Courtesy Johnson Motors.)

the hardest materials known, and is much harder than conventional cast iron. After the bores are machined, they are etched with acid to remove a few thousandths of an inch of aluminum. The silicon is unaffected by the acid and so the bore becomes almost pure silicon. The etch marks also serve as oil reservoirs for ring lubrication. It is said that the Vega block cannot be re-bored. General Motors does not provide over-sized pistons and rings. But hot rodders rebore them without much problem. This may be a preview of the future for small engines.

Whatever the type of bore, it should be checked for scratches, out-of-round, taper, and wear. The best way to do this is with an inside micrometer, although mechanics get a rough idea of the condition of the bore by the thickness of the ridge. This is the plateau at the top of the bore which marks the limit of ring travel. A cylinder more than .005" out of true should be rebored or discarded.

As previously mentioned, it is not practical to rebore and replate chromed cylinders. If there is evidence of excessive wear or if the chrome has started to peel, the unit must be replaced. Except for some motorcycles with very thin sleeves, cast iron can be bored oversize (for further information, see above under "Rings"). But before you begin work, make sure that oversized pistons and ring sets are available.

CONNECTING RODS

In small engines, conn rods are made of bronze, aluminum, or forged steel. The big end may be split (American practice) or one piece (English and European). The shank is usually an H section, although in some high rpm applications, tubular rods are used. The primary purpose of the rod is to transmit force from the piston—which can easily amount to several tons—to the crankpin. On four-cycles with full pressure lubrication, the rod also acts as an oil passageway to the wrist pin.

Single piece or full circle rods are found on many imported motorcycle engines and on variants of these designs used to power snowmobiles and go-karts. To remove this type of rod, it is first necessary to open the crankcase. The crankshaft must be forced apart with a press, and a dial indicator is used to align the crankshaft halves during assembly. Few mechanics will disturb these rods merely for inspection (Fig. 5-13). If the rod is obviously bent, or if the bearings are noticeably loose, the work is farmed out to a dealer.

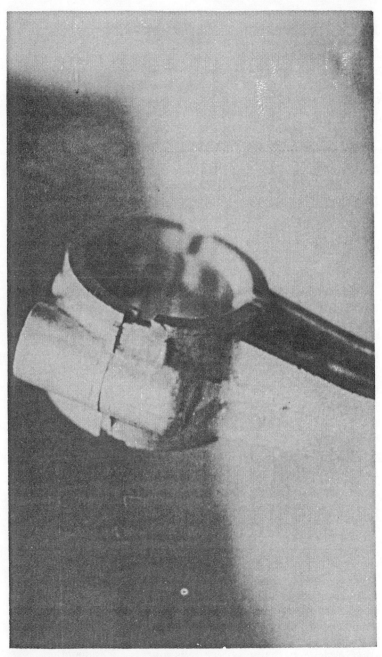

Fig. 5-14. Connecting Rod "Match Marks" in Alignment. All rods are coded but if you can't find the marks, scribe your own before the rod is disassembled. (Photo by Al Galinski.)

Split rods, on the other hand, are disassembled as part of every overhaul. Some rods have a right and a left side as defined (located) by an oil port and must be assembled in the original position. On multi-cylinder engines, the rod caps should be marked to prevent confusion. Theoretically, rods are interchangeable, but as a practical matter, each rod should go back into its own cylinder. In addition, each cap is matched to its own rod. If you look carefully you will see reference marks which must line up during assembly (Figs. 5-14 and 5-15). Torque to the manufacturers specs, and (on most engines) bend the locking tab against the heads of the conn rod bolts so that the bolts will not vibrate loose. It is good practice to replace conn rod bolts with each overhaul.

"Big End" Bearings

Many two-cycles and some of the better four-cycles use roller or needle bearings on the crankpin end (the big end). These are classed as anti-friction bearings and have the main advantage of being able to run with only a small amount of lubricant. On split rod applications, the bearings are uncaged, i.e., they are loose. Once the cap is removed, the bearings can fall into the crankcase. Be sure that you retrieve them all, since a single loose bearing can play havoc with an engine.

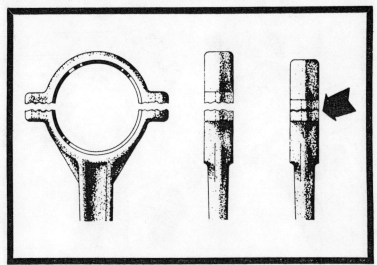

Fig. 5-15. McCulloch and one or two other manufacturers fracture the rod after it is finish machined. When assembled correctly, the fracture lines will almost disappear. (Courtesy McCulloch Corp.)

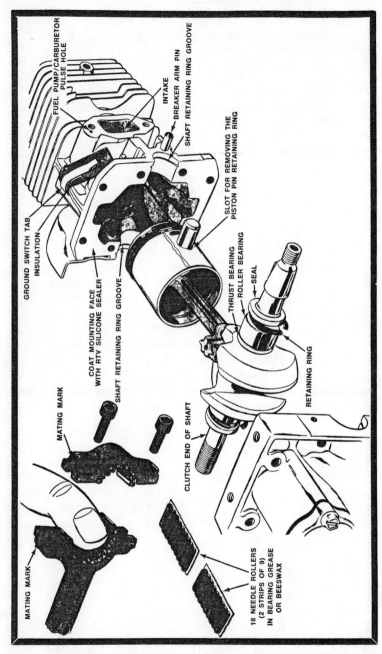

Fig. 5-16. Homelite Model 150 Chain Saw Engine. It is suggested that the crankcase fasteners be cleaned with Locquic grade T primer and secured with Loc-Tite Sealer. (Courtesy Homelite, A Textron Division.)

Bearings are replaced as a set, never singly. Coat the rod and crankpin with grease or beeswax to hold the bearings during assembly (Fig. 5-16). Bearings are replaced at over-haul or if there is any evidence of overheating (the bearings will have turned blue) or galling. A peculiarity of anti-friction bearings is that the tiniest speck of rust will cause rapid failure. Frequently this is seen in outboards which have been allowed to stand for a day or two after being dunked.

Plain bearings may be classified into three types according to structure. Older engines often employ poured babbitt bearings on the big end. Not many people work with babbitt any more, and if you should run across one of these engines, the best advice is to take it to a specialist. Often these men can be found doing maintenance on the equipment in large machine shops. Specify that grades SAE 10, B, or C be used.

Much more frequently encountered is the plain bearing that is integral to the rod. Wear means that the rod must be replaced, although an emergency remedy is to file the ends of the cap to reduce clearance. Lay a file on the bench; holding the rod cap at the center, move it against the teeth. Frequently check your work by bolting the rod on the crank. It must be said, though, that this "shady tree" fix is not recommended by any manufacturer.

Another type of plain bearing is the precision insert. These were developed to replace poured babbitt in automobile and aircraft applications, and have found their way into small engines. BSA has recently adopted inserts over the anti-friction type formerly used. Both Kohler and Wisconsin use them, as do some Honda models also. To replace, the old inserts are rolled out, the bearing backing surface cleaned, and new inserts installed. Some inserts have an upper and a lower half as defined by an oil hole.

Plain bearing clearances, when new, are in the range of .001"—.0015". Racing engines are set up a little looser with .002" or even more. The best way to check these clearances is to torque a rod to factory specs, measure the internal diameter with a micrometer and then measure the diameter of the crankpin in several places. Few mechanics do this however. A faster way, although it cannot determine crankshaft flatness, is to use Plastigage. This product is made by the Perfect Circle Corporation and is quite inexpensive. Place a quarter inch or so of the plastic strip on the high side of the crankpin at TDC. This is the area which gets most pounding. Now torque the cap to the recommended specifications. Without turning the crank, remove the cap.

You will find that the plastic strip has been flattened. This width, checked against the scale on the Plastigage package, tells you the bearing clearance. Lubricate before assembly.

"Small End" Bearings

As with the big end, these bearings may be anti-friction or plain. Normally the little end gives few problems and, if noise is not a factor, considerable wear can be tolerated. Sometimes you will find that a worn bushing has rotated in its boss and has cut off the oil supply to the wrist pin. The cure is either replace the bushing—which must be reamed to a push fit on the pin—or simply to anchor the old one with LocTite.

CRANKCASE

The primary function of the crankcase (or block) is to provide a steady bearing rest for the crankshaft. On four-cycles, the crankcase may serve as the oil sump; and on two-cycles, it is an integral part of the induction system. The crankcase also carries lugs for mounting the engine.

Cast iron cases, such as those found on industrial four-cycles, present no special difficulties for the mechanic. These cases are generally strong, and when properly aged, hold their dimensions. Aluminum cases are found on engines where light weight and good heat dissipation takes precedence over other design considerations. Because the metal is soft, threaded fasteners tend to wallow in their holes and strip out. One should be very careful not to over-torque these cases. If the threads show distortion, replace with the next oversize, or better, with a Heli-Coil or Tap-Loc inserts for "Group 2" metals.

Disassembly

Most aluminum cases are split vertically into two halves. The joint must be absolutely air-tight. Some designs employ a thin paper gasket between the two surfaces, while others depend entirely upon a metal-to-metal contact. It is imperative that the mechanic exercise extreme care in disassembling the cases. Remove the external flywheel and all fasteners, including the cylinder barrel. See that the exposed portion of the crankshaft is free of rust and scale, since the main bearing must slip over it. Expect the cases to stick a little—gasket varnish has been applied to the joint, and there are locating pins which sometimes are an interference fit in their holes (Fig. 5-17). If the engine has anti-friction bearing mains (and this includes nearly all motorcycle, outboard, and go-kart engines) it is often useful to heat the bearing boss. Heat will cause the aluminum to expand and free the bearing. Most mechanics prefer to use a hot plate for this operation, for excessive heat (as from a torch) can cause crankcase distortion.

Note: On most modern motorcycles, the transmission is in a unit with the crankcase. Exercise extreme care when separating the cases, so as not to disturb the transmission components.

Assembly

Assembly is the reverse of the above. Clean the surface carefully; if a gasket is used, replace it with a new one. Use a factory replacement because the thickness of the gasket is critical in terms of crankshaft endplay. If no gasket is used, coat the surfaces with a silicone cement such as Dow-Corning

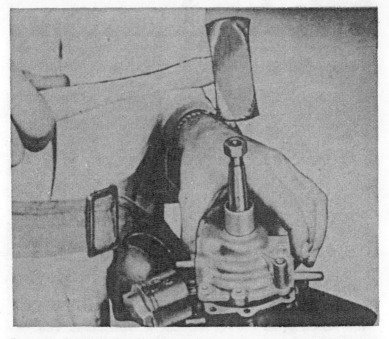

Fig. 5-17. Splitting the Crankcase Halves on a Lawnboy "A" Series engine. Note that the flywheel nut is used to protect the threads from the mallet. (Courtesy Outboard Marine Corporation.)

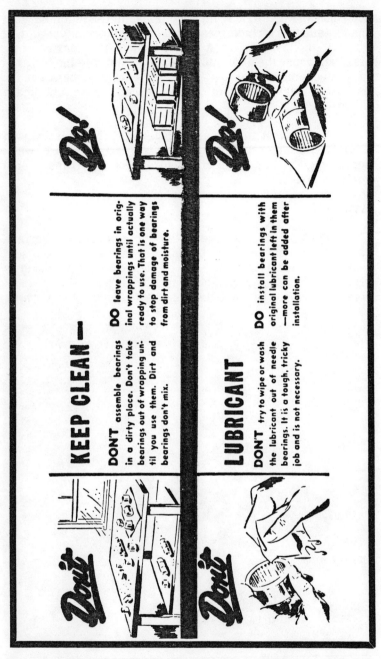

Fig. 5-18. Bearing Assembly Do's and Don'ts. (Courtesy McCulloch Corporation.)

"Glass and Ceramic Adhesive" or G.E. "Silicone Seal." Tighten the fasteners in sequence, from the center out to both ends. If the case sticks at some point, discover what is holding it. Do not use the fasteners to force it together. Heat on the main bearing retainer may be again necessary.

Bent cases are rare, but can occur from improper assembly, e.g., forcing the cases together, or from accident. When a rotary lawnmower blade contacts something solid, the crank and-or the case will bend. You can check the case with a known straight crankshaft. Binding on any point of the rotation of the crank means that the case must be discarded.

Pressure Testing

While few mechanics pressure test a crankcase, it is a technique which you should know about if you do much work on two-cycles. The test tool can be purchased from Burco Tools, Brandywine Station, Schenectady, N.Y., and consists of a bulb, a valve, rubber tubing, and a gauge. Assemble the crankcase with the shaft in place. Block off the intake and the cylinder register face with homemade plates and gaskets. The cylinder plate should have a fitting to accept the tube on the tester. Put 20 pounds on the case and watch the gauge. A new engine will hold pressure for an hour or more, a rebuilt unit for several minutes. A leak-down rate of no more than 0.5 lb a minute is acceptable. Immerse the assembly in water to find the leak. Generally, it will be the crankshaft seals, but the problem might be a porous casting. Often, repairs can be made with epoxy cement. Sometimes the leak will occur around a crankcase fastener, in which case the fastener can be sealed with a liberal coating of Loc-Tite Sealer.

Main Bearings

Plain bearings may be checked with Plastigage or with a micrometer (.001—.0015" is the clearance for new bearings). Ball and roller bearings must be checked for noise and for the presence of galling or discolored areas on the friction surfaces. When installing the race into its boss, it is a good idea to use a few drops of Loc-Tite to hold the bearing in place (see Fig. 5-18 for tips on servicing bearings).

Seals

All engines have some kind of seal around the ends of the crankshaft. These seals may be made of leather, but today are almost always neoprene. Oil seals may be removed by carefully deforming the metal retainer with a center punch and prying out with a screwdriver. Do not damage the area where the seal mates to the crankcase. It is possible to remove the seal without first removing the crankshaft, but one must be extremely careful not to nick the shaft. After the seal is removed, inspect the shaft for scoring, and polish lightly with emery cloth if necessary. A new seal is installed with the steep side of the lip toward the crankcase. This is so that crankcase pressure will tend to force the seal into closer contact around the shaft. The manufacturers name and seal number will always be on the side that you are working from. Lubricate the lips of the seal and spread a thin coat of Permatex Aviation No. 3 on the outside circumference of the retainer. Do not allow any of the compound to come into contact with the fiber portion of the seal. Slip the seal into position. If the crankshaft is in place, cover the keyway with masking tape. The edges are sharp and can cut the seal lip. Drive home with a wooden block or a piece of pipe. Pressure should be exerted on the outer edge of the retainer in order not to deform it.

CRANKSHAFT

This is the single most vital component of the engine. All too often mechanics lavish their time on secondary components such as pistons and valves and take the crankshaft for granted (Fig. 5-19). While it is still in the engine, take a sight along the power take-off end and crank. If there is any wobble, the crankshaft is bent and should be replaced or sent out to be straightened. Do not attempt to straighten the crank while it is still in the engine, as this procedure will almost surely damage the main bearings and-or the crankcase. Once the crank has been removed, you can check the bearing surfaces for wear and out of round as detailed previously. Light scoring on the bearing surfaces can be removed with emery cloth, but deeper scores mean that the crank must be machine ground. Many manufacturers supply main and big-end bearings in one or more undersizes. Others do not, and therefore make it mandatory that the crankshaft be replaced when seriously worn or scored. When ordering a new crank, be certain that you give the supplier all the pertinent data, including the engine serial number.

It should also be noted that bearing surfaces may be resized by metal spraying. The larger marine machine shops can do this work as can a few of the specialty automobile shops. In some instances the cost compares favorably with the cost of a replacement shaft.

On assembly, lubricate the crank with liberal quantities of engine oil.

Timing Marks

Four-cycle engines almost always have timing marks on the crank and the camshaft. Some engines built years ago did not have marks (possibly to make life difficult for non-factory technicians), and some engines are so designed that the marks wear away in service. The marks vary—they may be in the form of a circular indention, a punch mark, or a chamfered tooth. Lawson-Tecumseh engines use the keyway (Fig. 5-20) on the camshaft drive gear as an index. The crank and cam should be assembled so that the marks are in line (see Fig. 5-21) on nearly all engines. But sometimes due to manufacturing exigencies, the marks themselves are off. An example of this is the Tecumseh series built for Sears and equipped with the Craftsman suction lift carburetor. On these engines the correct timing is for the cam to be advanced one tooth to the left of the keyway reference mark on the crankshaft gear (Fig. 5-22).

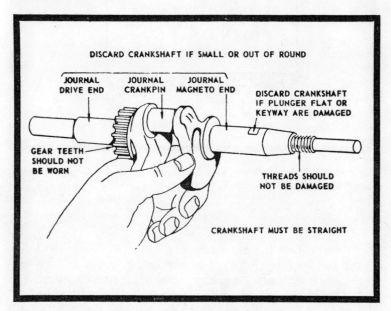

Fig. 5-19. A Check List for Crankshaft Inspection. (Courtesy Briggs and Stratton.)

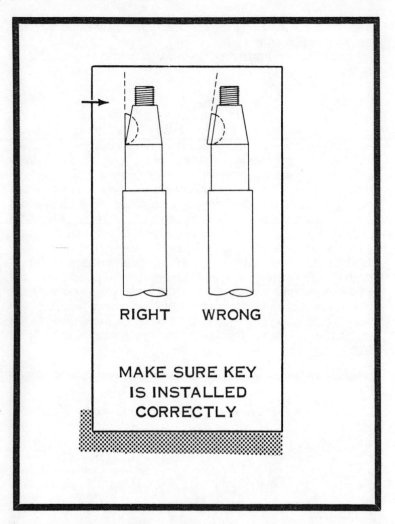

Fig. 5-20. The Correct Way to Install a Half-Moon Key on a tapered shaft. (Courtesy Outboard Marine Corporation.)

Camshafts

Rarely does the cam give problems. But once the lobes begin to wear past the case hardening, rapid failure will result. Bearings are generally replaceable and camshaft end play should be .001" to .002". On assembly, lubricate the cam with engine assembly oil, or better, with high pressure cam lube available from automobile specialty outlets.

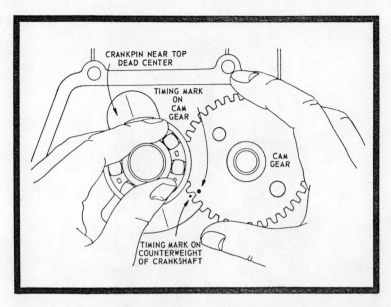

Fig. 5-21. Position of the Timing Marks on Briggs Ball-Bearing Engines. On the plain bearing models, the crankshaft mark is on the gear. (Courtesy Briggs and Stratton.)

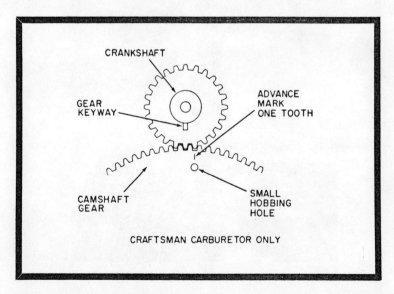

Fig. 5-22. An Exception to the Practice of Aligning Timing Marks: Tecumseh Engines with Craftsman Suction-Lift Carburetors. (Courtesy Tecumseh Products Company.)

DRILLING

TAPPING

C

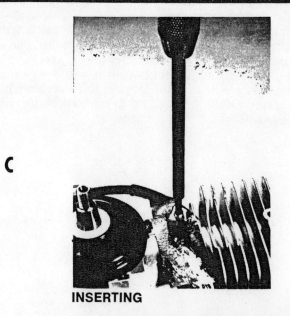

INSERTING

D

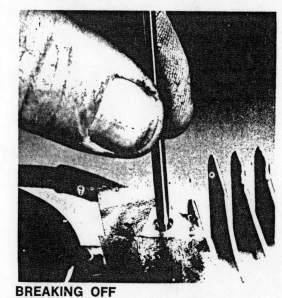

BREAKING OFF

Fig. 5-23. "Re-Sizing" Spark Plug Hole for Heli-Coil Insert. Note masking paper to prevent chips from entering chamber or entering heat dissipation fins. (Courtesy of Homelite, a Textron Division.)

INSERTING A HELI-COIL

Oftentimes, a careless person will have snapped a spark-plug off. Provided you can get a grip on the remains, you can usually unthread it, clean the hole, and put in another plug. But if you can't turn it, you must drill it out. This usually results in an oversize hole. After tapping this oversized hole, you should insert a Heli-Coil sized for the new (oversize) plug. Fig. 5-23 shows how.

Chapter 6
Power Transmission

Power can be transmitted from an engine in various ways: by shafts, friction discs, V-belts, chains, and gears. As a mechanic, you will spend as much as one quarter of your time servicing power trains. It is necessary to have some understanding of the principles involved.

SHAFT DRIVE

The most frequently used method of power transmission, shaft drives are found, in one form or another, on all small engine applications. Shafts in combination with gears are used on outboard engines, motorcycles, and, with V-belts and chains, on various types of garden and recreational equipment. Direct shaft drive is used on most rotary lawnmowers, that is, the blade is fastened directly on the end of the crankshaft. In some designs, the blade is mounted by means of an aluminum adapter which is keyed to the crankshaft. The adapter has two projections which fit into holes in the blade. These projections are relatively weak, and will shear if the blade hits something solid.

Lawnmower Shaft Inspection

When servicing a rotary lawnmower, the first thing to do is to check the straightness of the shaft and the blade. This is done by removing the spark plug lead and clipping it to one of the cooling fins so that it cannot touch the plug terminal. Turn the mower on its end—with the cylinder head up to prevent oil flooding the chamber—and slowly turn the blade by hand. See Fig. 6-1. Take a sight on a reference point under the deck and see that both ends of the blade register within 1-16 of an inch of each other. If the blade appears to be bent, remove it and check the shaft. One bent part usually means that both are damaged. The quickest way to check the shaft is to "eyeball" the hole in the center while someone spins the motor. Wobble means that the shaft is bent.

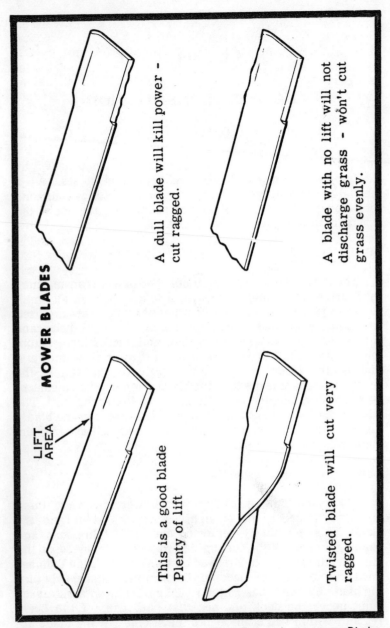

Fig. 6-1. Various Maladies which can affect Rotary Lawnmower Blades. If the blade is bent or ground unevenly so that it is out of balance, the mower will vibrate, causing engine damage. An inexpensive static balancer can be purchased from lawnmower parts outlets or from Sears. (Courtesy Outboard Marine Corp.)

A bent shaft should be replaced. This can be relatively expensive since it involves a complete teardown and the purchase of a new part which can cost anywhere from twelve to twenty-five dollars. An alternative method is to straighten the shaft. Small engine manufacturers agree that this should not be done, but if the shaft is straightened correctly it will usually perform as well as a new one. Some dishonest shops straighten the old shaft and "sell" the customer a "new" shaft.

Straightening a Shaft

The correct procedure is to remove the shaft from the engine and straighten it with a press and a dial indicator. Usually the bend will be on the shaft extension and will radius out from the lower main bearing. Sometimes the shaft will have bent at the crankpin and must be discarded. Before assembly, the shaft should be sent out for Magnaflux or Zy-Glo testing. Either of these processes will detect surface cracks too small for the naked eye to see. The testing is very inexpensive.

Some shops straighten shafts without removing them from the engine. They use a hydraulic jack and a support which loops around the machine. This method is not recommended.

Blade-to-shaft adapters can also bend and can wear in the keyway if they have not been tightened down. A worn adapter will make a distinct knocking noise at low speeds.

What has been said for rotary lawnmowers shafts, goes for all types: inspect every shaft for bending. Short shafts can be chucked up in a drill press, and the longer ones, such as outboard engine driveshafts, can be rolled on a flat surface or turned in a lathe. Other indications of bending are worn or egg-shaped bearings, and excessive vibration.

Shafts used on reel mowers and home garden equipment are usually hardware store items. If the shaft is bent or worn, it can be replaced with stock of the same diameter. These shafts are formed by drawing the steel bars through dies. The compression gives a fairly hard surface to the shaft and makes it suitable for light duty use in bushings. On most of these applications, the steel used is SAE 1112 or 1120. Twice the strength can be had if you substitute SAE 3140, 2340, or 6140.

FRICTION DRIVE

These drives are found on a few motorbikes, such as the French Solex, and are incorporated as a safety feature on

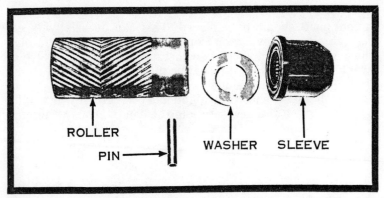

Fig. 6-2. Friction Roller as used on Power Lawnmowers. After long service, these rollers require replacement. (Courtesy Outboard Marine Corp.)

some outboard props and lawnmower blades. If a solid object is struck, the drive will slip, thus protecting the driveshaft. The most common use of friction drive is on garden tractors and on the propulsion mechanism of rotary lawnmowers (Fig. 6-2). Drive rollers should be replaced when worn (most are secured by 1/8" compression pins). For long life, the driven wheels should be the same height as the front wheels.

V-BELT DRIVE

V-belts are widely used in lightweight equipment, not over 20 HP. Operation is silent, a simple idler wheel can serve as a clutch, the ratios can be varied by changing the diameter of either pulley, and rotation can be reversed by crossing the belt. V-belts also protect the drive mechanism by cushioning the shock of uneven loads.

The belts are constructed of textile or synthetic cords and are covered by fabric impregnated with rubber. Heavy duty belts may have one or more steel cords which act as the load-carrying member. There are many specialized types of belt available for different environments. If you have a problem with excessive belt wear, it is certainly worthwhile to call one of the major manufacturers. B.F. Goodrich claims that their new Glasstex design outwears conventional belts by at least 4 to 1 and, in some applications, by a factor of 10 to 1.

ADJUSTMENT AND REPLACEMENT

However, belts do require periodic adjustment and replacement. The condition of the belt can tell you much about the machine. If it is burned in spots, the belt was not adjusted properly, or the machine has been overloaded. Wear on one side means that the pulleys are slightly out of alignment. See Fig. 6-3 for proper installation. A belt that is twisted inside-out indicates extreme misalignment and should be replaced, since one or more of the cords has broken. Wear on the reverse side might mean a frozen idler or misadjusted belt stops. Checking or cracking on the inside of the belt is a result of overheating. Oil indicates a leaking seal, which in turn might be caused by a bent shaft.

Replacement belts should exactly match the original in width, length, and type. The correct size can be determined by comparison (expect the old belt to have stretched a half inch or so), or else by the code number stamped on the back of the belt. This may be a part number which the manufacturer of the machine has put on the belt to discourage "off brand" replacements, or it may be a code number indicating the belt size. Unfortunately there isn't much standardization of code numbers among belt manufacturers. The Gates Truflex is often encountered. The first digit refers to the cross-section size of the belt: No. 1 is ⅜", and No. 2 is ½". The second two digits refer to the outside circumference in inches, and the last digit indicates additional tenths of an inch in outside circumference. For example, a Truflex marked 2200 is a ⅜'s inch belt, 20 inches in circumference. Because of the variations in coding, every shop should have a belt measuring stick. B.F. Goodrich supplies this tool as stock no. MSG. Goodrich also supplies a 76-page replacement guide which includes cross reference data, belt popularity tables and numerical listings (stock no. IPC-164-7).

Belt Tension

Belt tension is important. Loose belts cause slippage, heat buildup, loss of power, and squeal. Whenever a belt squeals, it

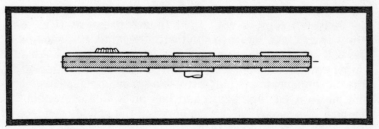

Fig. 6-3. Sheaves and Idler showing correct (straight-line) relationship. On other designs, the idler may ride on the outside surface of the belt. (Courtesy Outboard Marine Corp.)

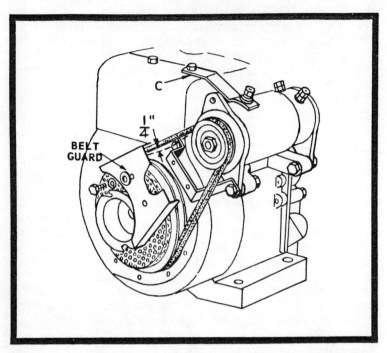

Fig. 6-4. Proper Belt Adjustment. Use light pressure at a point mid-way between the sheaves. (Courtesy Briggs and Stratton.)

is slipping. Generally the problem is inadequate tension, although it can be caused by worn sheaves (pulleys) or momentary overloads. Do not attempt to fix a slipping belt by using belt dressing. These preparations will stop slippage, but they contain chemicals which are harmful to rubber compounds.

Too much tension will stretch the belt, cause early bearing failure, rapid sheave wear, and overheating. There are several ways to adjust the tension of V-belts. Perhaps the most common is to adjust the belt so that it deflects between ¼" and ½" between centers with light thumb pressure (Fig. 6-4). Another way is to strike the belt with the hand. If it's too tight, the belt will feel solid, and if too loose it will feel "dead." Under correct tension, the belt will be "alive" and "springy." The most accurate method is to use a spring gauge. This tool is similar to a postage scale and has an indicator reading in pounds. It is hooked under the belt and the pounds per inch of deflection is read directly.

Whenever belts are replaced, inspect the sheaves. A new belt should be level with or no lower than 1/16" below the top of the sheave. Under extreme wear the belt can drop down to the bottom of the groove. In this condition, little power is transmitted since a V-belt depends on the wedging action of its sides against the sheave walls. In dusty conditions, such as encountered with lawn edgers and garden cultivators, the sheave will wear more rapidly than the belt. Sometimes it is helpful to replace the original equipment aluminum sheaves with cast iron or cast steel. However, rust can be as damaging as wear. The inner surfaces of the sheave are highly polished and so are very susceptible to rusting.

It is important that the sheave be aligned as accurately as possible with its mate and that both run true. A bent sheave or one that is running on a bent shaft will quickly destroy itself and the belt.

CHAIN DRIVE

Found on a wide variety of power equipment, chains are almost universally used to transmit power to the back wheel of motorcycles. Chains are dirty, messy to service, and require periodic replacement. Motorcyclists sometimes wonder why their machines are fitted with this seemingly archaic form of power transfer. With few exceptions (such as the current GM front drives) chains disappeared from automobiles fifty years ago with the development of the Hotchkiss and torque tube shaft drives.

Why then are chain devices still used on so many motorcycles? Economy is part of the answer. To transmit power from an engine located amidships in the frame to the back wheel would require a series of gears or else a shaft and assorted bevel gears. If the engine is mounted with the crankshaft running longitudinally (fore and aft) in the frame, the shaft will require one bevel gear to transmit motion to the back wheel. Both BMW and Moto Guzzi use an inline engine and a single bevel. If the crankshaft were mounted across the frame (transversely), as is common motorcycle practice, two bevel gear sets would be required—one to make a 90 degree turn coming out of the transmission and another to turn the drive at the rear wheel. In either case, the shaft would have to be offset enough to pass by the wheel and, if a spring suspension were fitted, the shaft would need to have a U-joint. Gears, U-joints, bearings, and the rest of the hardware means additional weight and extra expense.

But modern motorcycles can cost on the far side of three thousand dollars, and expense is not an overriding consideration. Complexity reduces reliability, and added hardware requires additional power and complicates periodic maintenance.

There are other arguments for staying with chains. One of the most compelling is the matter of efficiency. A typical motorcycle transmission will absorb from 5 to 7 percent of the input energy. If the engine produces 20 horsepower on the input shaft, only 19 HP will get through the gear box and to the output shaft. The other "horse" is used to turn the gears and to churn the oil. It is converted into heat which passes to the atmosphere through the transmission case. We cannot do much about transmission losses (although at least one vintage English motorcycle had a transmission consisting of chains and dog cluches—it was efficient, but bulky). In contrast, chains only lose about 2 percent of the input energy. Another way of saying this is that a motorcycle gear box is between 93 to 95 percent efficient and a drive chain is 98 percent efficient. Every "pony" counts.

Chains also tend to cushion the drive line by absorbing shocks and some of the braking and acceleration loads.

Drive Ratios

If the customer wants to vary the drive ratio, this can be easily done by changing the sprockets. If the drive sprocket has 20 teeth and the driven sprocket has 60, then the overall reduction is 3 to 1. To change the reduction to 2 to 1, all that is necessary is to install a 30 tooth drive sprocket. If the customer wants a greater reduction, the size of the large (driven) sprocket can be increased. The reason that we do not employ a smaller drive sprocket is that the chain does not move over the sprocket evenly: it is subject to a cyclic speed variation. This is because the sprocket is not round, but is actually (from the point of view of the chain) a polygon. This irregularity has little affect on large sprockets but becomes serious with the smaller ones. For this reason, a 19-tooth drive sprocket is the absolute minimum. Wherever possible, use larger sprockets.

Chain Stretch

Sometimes a customer will want a heavier chain than the one originally installed. This modification is usually wasted time and money. The ⅜" chain used on lightweight motorcycles has a breaking strength of 1 ton, which is far and above the power developed by the engine. The ½" chain will tolerate a force of 2 tons, and the ¾" variety a force of 3¼ tons. In most cases, the customer will have mistaken chain stretch for evidence of impending failure. All chains, except the special Reynold's pre-stretched racing chain, grow in use. This is perfectly normal.

Chain Wear

The big problem is chain wear. Once the surface hardening on the pins and link plates is gone, the chain will literally saw itself to pieces. Harley-Davidson suggests that the chain be squeezed together a few links at a time and then stretched. If the play is more than 1", the chain should be replaced. Another test is to secure one end of the chain and stretch sections of it.

10 inches on a ⅝" pitch chain equals 16 pitches. Maximum allowable length is 11¾".

9 inches on a ⅜" pitch equals 24 pitches. Maximum allowable length is 9 3-16".

Inspect the sprockets for wear and for hooked teeth. All sprockets should be in line.

Periodic Maintenance

Chains should be periodically cleaned in solvent or, if rusted, in caustic. Numerous lubricants are on the market, including some with fairly exotic ingredients such as molybdenum disulfide. A few enthusiasts dip their motorcycle chains in a heated mixture of paraffin and chassis lube. The chain is hung up to dry and the excess wiped off. But the major concern in chain maintenance is to keep it clean and to lubricate it often, if only with motor oil.

Chains should be set-up with some slack. An overly tight chain will whine and a loose one will "snatch" and may come off the sprocket. On motorcycles, the average adjustment is between ½" to ¾" play, measured at a point half-way between the sprockets with the machine on its wheels and loaded. The closed end of the master link lock should point in the direction of chain movement.

GEAR DRIVES

Most gear drives encountered in small engine work are reduction gears, i.e., they are designed to reduce the speed of the output shaft. The gears are generally straight-cut, that is, the shaft bore and teeth are in parallel. Since the leading edge of the tooth makes full contact, these spur gears are very

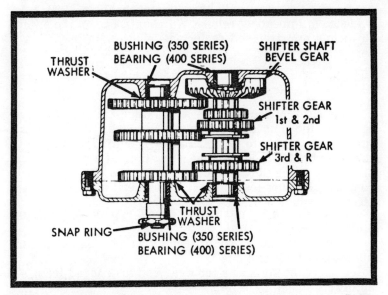

Fig. 6-5. Peerless Series 350 Sliding-Gear Transmission used on Riding Mowers and Garden Tractors. The shifter gears turn with the input shaft and move along the shaft to engage one or the other output gears. The reverse idler gear is not shown. (Courtesy Tecumseh Products Co.)

strong, but are noisier than the curved or helical gears used in automobile transmissions and on the primary side of a few Japanese motorcycles. When setting these boxes up, some end-play is desirable, with .003" as a good average figure. Adjustment is made by the number of gaskets used on the gear box cover or by the thickness of the thrust washer. Lubrication requirements vary with the manufacturer. On-engine units often use motor oil, and may share the sump with the engine, while remote units often take EP 90 grease. (EP means Extreme Pressure.)

Transmissions

Small engine transmissions are normally not dismantled for inspection. The only time the unit should be disturbed is during major overhauls or when there is evidence that the unit has failed, or is about to fail. Excessive noise, refusal to accept or to remain in a gear, or metal particles in the oil, are reasons for a teardown.

Two types of transmissions are used: the sliding-gear type (Fig. 6-5) in which the gears themselves are moved in and out of mesh, and the constant-mesh type. The shifting gears are positioned on a shaft by means of a key and move laterally along the shaft. One side of the shifting gears and the driven gears are beveled to make engagement a little easier. Constant mesh transmissions (Fig. 6-6) are used on some riding lawnmowers and almost universally on motorcycles. All gears turn, but only the engaged gear is pinned to rotate the output sprocket. Fig. 6-7 is a drawing of a Peerless Series 200 two-speed transmission. The shifter lug (No. 9) is pinned by the key (No. 12) to the input shaft (No. 7). The input gears (No. 5 and No. 6) are free to spin on the input shaft; the output gears are pinned to their shaft. No power is transmitted unless the shifter lug engages one of the input gears. When it does the gear in question rotates and turns the output shaft. The Albion 3-speed box in Fig. 6-8 is similar in operation.

Inspection Before Overhaul

Before a transmission is taken apart, check the clutch for full release and engagement. Remove all linkages and attempt to shift the box by hand. Often the problem can be cured by simple adjustment. It is impossible to be specific in a book of this type, but worn detents, bent cams, loose or sheared pins—in short, any waste motion in the shifter mechanism—can cause what appears to be transmission failure. Refer to Fig. 6-8 for the detent type shifters used on English bikes and to Fig. 6-9 for the BMW variety of cam shifts.

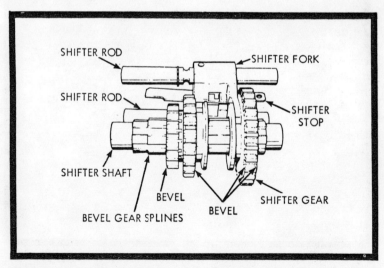

Fig. 6-6. The Shifter Mechanism of the Transmission in Fig. 6-5. (Courtesy Tecumseh Products Co.)

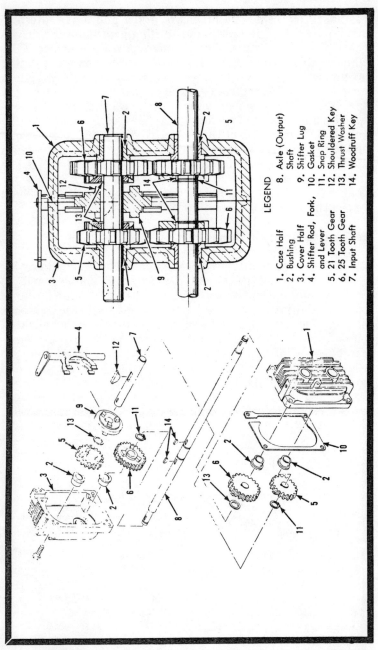

Fig. 6-7. A Constant Mesh Two-Speed Transmission (Peerless 200). The more complicated 4, 5, and 6-speed units operate on the same principle. (Courtesy Tecumseh Products Co.)

Overhaul

If the transmission has to be taken apart, drain the oil and clean the housing with solvent or Gunk. Note the length of the various bolts and screws. Being careful not to disturb the gears, lift off the cover. At this point, you will have a good view of the internal parts of the transmission. Study it; time spent in understanding the box will not be wasted.

Remove one part at a time and place it on a clean bench in the order of removal. You can usually assume that the box is assembled as it came from the factory, since "shady tree" mechanics have a healthy respect for transmissions.

Observe the position of the spacers. These hardened-steel washers are used to protect soft metal from contact with the gears and help set the internal clearance of the transmission. Generally, the oil seals and O-rings are replaced when the transmission is dismantled. In some designs, the thickness of the cover gasket is critical. Inspect the shift fork for wear and bending. If the unit has two shift forks, be sure to note which one goes where. The right fork on small Hondas is stamped with an "R". Examine the gear teeth for chipping and for uneven wear patterns. The latter would indicate bent shafts and-or loose bearings. The shaft can be chucked up in a drill press to check for trueness. The splines should be smooth, without pits or burrs.

As a rule, the shafts turn on bushings or in caged anti-friction bearings. An exception is the American-made Harley which uses scores of uncaged (loose) rollers. Each of these rollers must be accounted for, and rollers from different sets cannot be mixed. If a single roller is lost, the whole set must be purchased, since a new roller will have a slightly greater diameter than the used ones. Check bushings for play. Anti-friction bearings should be spun by hand (do not use compressed air) to determine if they are noisy. Blind bearings, that is, bearings which are pressed into closed bosses, so that you can only work from one side, can be removed with special pullers or by hydraulic force. To remove a bushing, use a shaft exactly the size of the bushing bore. Fill the bore with heavy grease and strike the shaft with a hammer. The bushing will be rammed free. On needle bearings, the procedure is the same, except use oil-soaked newspaper as the medium. In open holes where it is possible to get behind the bearing, you can use an arbor press or a drift. Sometimes it is helpful to heat the case in the vicinity of the boss with a propane torch. Once a pressed-on bearing is removed, it should be discarded. New bearings are installed from the numbered side to the

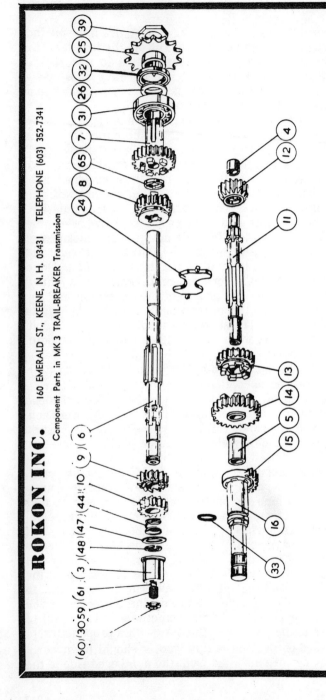

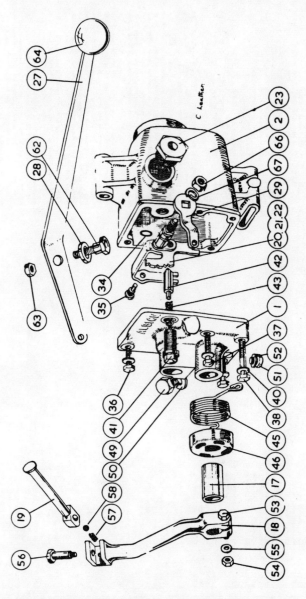

Fig. 6-8. The Albion Transmission used on the Rokon MK. 3 Motorcycle. Part No. 15 is the kickstarter segment. (Courtesy Rokon, Inc.)

PARTS LIST FOR FIG. 6-8

Ref. No.	Part No.	Description
	020	Complete Transmission
1	EJ 1	Cover
2	EJ 2/P	Case - pivot fixing
3	EJ 13	Mainshaft adjuster bush
4	EJ 7	Layshaft box bush
5	EJ 30	Layshaft bush - inner segment
6	EJ 3	Mainshaft (Special, Rolon)
7	EJ 4	Mainshaft high gear pinion 22.T.
8	EJ 8	Mainshaft sliding gear 18.T.
9	EJ 28	Mainshaft low gear pinion 13.T.
10	EJ29/1	Kickstart ratchet pinion 14.T.
11	EJ 6	Layshaft
12	EJ 5	Layshaft high gear pinion 14.T.
13	EJ 9	Layshaft sliding gear 18.T.
14	EJ 10	Layshaft low gear pinion 23.T.
15	EJ 27/1	Kickstart segment
16	EJ 34	Kickstart shaft - serrated
17	EJ 34a	Kickstart shaft distance tube
18	CJ 8	Kickstart crank
19		Kickstart crank - pedal
20	EJ 11/12	Inside operator - bow (see note #5)
21	EJ 11/12	Inside operator - shaft (see note #5)
22	EJ 11/12	Inside operator - shaft rivet (see note #5)
23	EJ 14	Operator fork
24	EJ 16	Operator lever
25	BJ 126	Final drive sprocket 14.T. ½" x 305"
26		Sprocket spacer
27	BJ 11	Control lever
28		Control lever spacer
29	EJ 47c	Side lever
30		Hardened pad (see note #6)
31	A 5	Ballrace 20 x 47 x 14 m/m 120.V2
32	A 35, 1	Oilseal 15/8" x 15/32" x 3/16"
33		O-Ring 5/8"
34		O-Ring R. 2068/P.S.
35	EJ 13	Anchor pin ½" x 20 x 9/16" hex head
36	EJ 15	Cover pin ½" x 20 x 1" hex head
37	EJ 23	Cover pin ¼" x 20 x 1" hex head (3 per transmission)
38	EJ 23	Cover pin ¼" x 20 x 13/16" hex head
39	A 14	Sprocket lock nut
40		Washer ¼" dia. (5 per transmission)
41	BJ 16	Plunger box
42	BJ 18	Plunger spring
43	BJ 17	Kickstart ratchet spring
44	CJ 15	Kickstart return spring
45	CJ 7	Kickstart return spring cover
46	CJ 6	Mainshaft washer
47	C 124	Split ring
48	C 124	Filler plug 7 16" whit
49	BJ 23	Fibre washer 11 16 x 29 64" x 5/64"
50	BJ 23a	
51	A 24	Drain plug 5/16" whit
52	A 24a	Fibre washer 9/16" x 5/16" x .080"
53	C 115	Kickstart crank pin nut
54	C 115	Crank pin washer 5/8" x 5/16" x 1/16"
55	C 115	Crank pedal pin (see note #7)
57		Crank pedal spring (see note #7)
58		Crank pedal ball ½" dia. (see note #7)
59	C 132	Barrel adjuster (see note #6)
60	C 133	Barrel adjuster nut
61		Ball 17/64" dia. (see note #6)
62	EJ 35	Shoulder pin
63		Nut 5/16" x 26
64	G 91a	Knob - black 1 1/2" dia.
65	EJ 36	Thrust washer
66	EJ 48	Operator nut 3/8" x 26 x 5/16"
67	EJ 48	Washer 3/8" single coil

NOTES:

1. Gearbox MOH. 90 to MOH. 548 are fitted with a 13-tooth Kick starter pinion (stub tooth) in lieu of the 14-tooth (involute tooth). These pinions are interchangeable with each other.

2. Gearbox MOH. 549 and onwards are fitted with a 14-tooth kick starter pinion (stub tooth) and a 27-tooth kick starter segment; these gears supersede all previous type ratchet pinions and segments which are now obsolete and for all orders received for either of the latter a pair of the new gears will be supplied.
When replacing old type gears it may be necessary to scrape cover to provide extra clearance for the ratchet pinion which has a slightly larger outside diameter.

3. Gearbox MOT. 1 and onwards are fitted with an externally fitted rubber oil seal in lieu of the internally fitted felt oil seal, A. 35 which dispenses with the use of oil seal retainer A. 36. In future, gearbox case EJ.2 will be supplied complete with rubber oil seal and in this state it is interchangeable with old type, therefore the existing part number will be retained. As the new oil seal on its own is not interchangeable with felt seal and retainer, the latter will still be available as spare parts.

4. Gearbox MOT. 1 and onwards are fitted with a modified final drive sprocket and this is not interchangeable with the old type unless modification number three is used.

5. Gearbox MPH 110 and onwards, fitted with modified inside operator and shaft assembly EJ 11/12 and inside operator shaft bush EJ 14. Diameter of shaft increased from ½" to 5/8" to facilitate fitting of oil seal ring - EJ 14 bush is still available but inside operator with ½" shaft is obsolete and must be replaced with new pattern operator and bush.

6. Part Nos. 5, 30, 59 and 61 are sold as one unit.

7. Complete kick start lever consists of part Nos. 18, 19, 53, 54, 55, 56, 57 and 58

exact depth of the original. Larger bearings may have the outer race pressed on to the cover and the inner race pressed on to a shaft. First install the outer race, applying force to the race only. Then, supporting the inner race with a hollow tube, press on the shaft. If the boss has score marks, showing that the bearing has turned in it, use Loc-Tite to secure the new bearing.

Assemble the transmission in the sequence in which it came apart. Nothing should be forced. On sliding gear types, the bevel side of the teeth should face each other. Install the cover, making sure that the locating pins line up with their holes.

Test the unit by turning the input by hand and counting the revolutions of the output shaft in each successive gear.

Outboard Gear Cases

The gear cases on outboard engines contain a bevel gear to transmit the power from the shaft through a 90 degree turn.

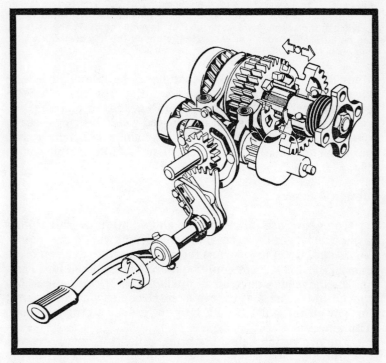

Fig. 6-9. The current BMW Transmission. Note the cam type shifter. (Courtesy Butler & Smith, Inc., Distributors for BMW.)

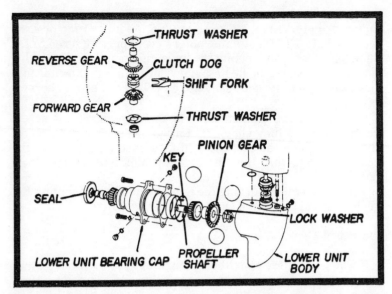

Fig. 6-10. The Lower Unit of the "Sport Scott" 27 HP engine. When correctly positioned, the clutch dog should be exactly mid-way between forward and reverse gears with the shift lever in neutral. The shift rod can be adjusted through an inspection plate on the low port side of the unit. (Courtesy McCulloch Corp.)

Many outboards include a second bevel (on the other side of the shaft) which is the reverse gear. Fig. 6-10 shows a Scott gear case with forward and reverse gears. In addition, the gear case will contain ball or tapered-roller bearings, seals, spacers, a propellor shaft, and, on engines so fitted, a shift mechanism. Usually this mechanism is quite simple, as in Fig. 6-10.

Design Factors

The gear case must be light and small in overall dimensions. The larger the case, the less efficient will be the engine. Years ago, when it was rare for an outboard to develop more than 25 HP, gear cases could be more or less taken for granted. Engines today develop as much as 120 HP and turn 5,000 rpm or more. The gear case must be strong enough to transmit this power and must have a built-in reserve strength to cope with accident.

The protrusion below the prop, called the skeg, protects the propellor and supports the engine when it is out of the water. Above the propellor is an anti-cavitation plate which should run parallel to, and just below, the surface of the water.

Some engines have a tab mounted on the anti-cavitation plate, aft of the propellor. This tab is analogous to the trim tab on an airplane. The lower part of the prop operates in less-disturbed water than the upper part. The lower part has a better "push" and tends to move the stern of the boat in the direction opposite to prop rotation. A prop turning counterclockwise will move the stern to port (left). A clockwise prop will nudge the boat to starboard. The tab cancels these forces. Some makers mount the gear case at a slight angle to produce the same effect. Twin engine units normally rotate in opposite directions and cancel each other.

The exhaust is routed through the lower unit and is expelled in the disturbed water behind the propellor. Cooling water from the powerhead is mixed with the exhaust to prevent overheating of the lower unit.

Maintenance

The gear case requires very little service, other than routine maintenance. Most manufacturers suggest that the lubricant be changed every 100 hours of operation or once a year, whichever is sooner. Only recommended lubricant should be used and these recommendations vary. For example, Chrysler warns against using extreme pressure hypoid type grease, while McCulloch says that only EP Hypoid 90 be used. Check the handbook for each type of engine. To change the lubricant, mount the engine upright and remove the drain plug (at the bottom of the lower unit), the vent plug, and the filler plug. The lubricant should just come up to the level of the vent plug. Replace the plugs, using new gaskets if necessary, and tighten securely.

Probable Faults

Several problems can occur with the gear case. If the seals are bad, water will get into the lubricant and cause rusting. Any signs of water mean an immediate teardown, since the seals must be replaced and it can be assumed that the bearings are damaged. After long service, the bearing and gears will develop excessive play which will cause accelerated wear. Unfortunately, it is next to impossible to hear a worn lower unit. There is so much other noise that gear and bearing whine are masked. But wear shows up in side-to-side and fore-and-aft play of the propeller shaft. A new shaft will have no more than .001" of fore and aft movement and zero side play. If a bearing fails, it is not unknown for the case to break. Do not attempt to repair it, since the case will certainly be distorted. These cases are made of thin, cast aluminum and are strong only as long as their structure remains integral. Some shops will heli-arc a broken case, but the practice is highly dubious.

Other than water contamination, the most frequent service problem is a refusal to shift. Nearly all units have a shifter mechanism similar to the one shown in Fig. 6-10. The disposition of parts may be slightly different in other makes. For example Chrysler, OMC, Mercury, and West Bend have the forward and reverse gears mounted on the propellor shaft, rather than on the drive shaft. But the principles are the same: a fork moves a dog, which is keyed to revolve with the shaft, to engage one or the other gear. A refusal to shift can mean a bent shift fork (caused by forcing a gear), a misadjusted linkage, or a sticking linkage. Whenever possible, disconnect the linkage and try to engage the gears by hand. If the gears engage, you can be sure that the problem is in the linkage and not in the gear case. Adjustment of the linkage varies with the make and model of the engine. On two-cylinder Chryslers, the shift rod is located on the starboard side of the lower unit and has an adjusting nut which varies the length of the rod. This nut should be adjusted so that there is an equal throw (movement of the rod) for Forward, Neutral, and Reverse. Essentially the same procedure is used on the McCulloch Custom Flying Scot. On the OMC manual shift models, Forward and Reverse should engage on high points of the shifter lock and equally distant from the Neutral detent. The early production West Bend had the gear shift control rod and the shifter rod—the one going into the gear case—mounted on a single bell crank. Correct adjustment required that the shift control rod be horizontal with the gear box in Neutral. Late model West Bends have a layout similar to the Chrysler.

Overhaul

As with any transmission work, extreme care and cleanliness is required. The bench should be clean and covered with newspapers. As each part is removed, it should be layed out in sequence to make reassembly easier. On many engines, the reverse and forward gears are identical—but used gears must be replaced in the position in which they were found. The same is true of spacers and bearings.

Disassembly is usually quite straightforward with few complications such as the need for special tools. An exception is the special tool required to remove the shift dog on Mc-

Culloch engines. This is spring compressor part number J7534 available from Kent-Moore Organization, Inc., 28635 Mound Road, Warren, Michigan. On Mercurys, the forward and reverse gear may have to be removed from the housing with a slide hammer puller of the type used to remove starter motor bushings in some automobiles. Some mechanics prefer to use a small amount of heat and a wooden block to jar the gears loose.

Once the transmission is disassembled, each part should be checked carefully. The prop shaft can be chucked up in a lathe to verify its straightness.

Note that 1967 and later OMC manually shifted engines have a spring and ball assembly under the clutch dog. When removing the dog from the shaft, be careful not to lose the balls. The dog has indentations milled on the inner bore for the balls and they must be positioned correctly when the dog is assembled back on the shaft.

Each part should be checked for trueness and unusual wear patterns which would indicate problems in alignment. Splines may have been damaged if the prop struck a solid object. All gasket surfaces must be cleaned and inspected for nicks. The bearings should be spun by hand—never use compressed air—to detect noise and roughness. Replace any suspect bearing.

Reassembly

Assembly is the reverse of disassembly. Use new gaskets and sealer such as OMC No. 1000. Use new O-rings and seals. The propellor shaft seal is installed with the lip toward the prop in order to keep water out of the unit. The drive shaft seal has the lip pointed downward to keep lubricant in the gear case. Oil the assembly lightly to prevent damage to O-rings and seals, but do not fill the unit until the backlash has been adjusted. (Some engines have a fixed backlash.)

Adjusting Backlash

Backlash is the most critical aspect of servicing lower units. If a mistake is made, the unit will quickly wear and can even destroy itself. To understand the importance of backlash, look at Fig. 6-10. The fore and aft movement of the pinion gear controls the depth of contact with the teeth on the forward and reverse gears. Too deep a mesh will cause friction and heat; too shallow a mesh will not provide enough metal-to-metal contact to transmit the power. A related problem is fore and aft play on the shaft between its bearings. These adjustments are made by adding or subtracting shims.

There are two broad approaches to adjusting backlash. The best method, and the one which all beginners should use, is to purchase a factory manual for the particular engine and to follow the instructions religiously. Sometimes, special jigs and fixtures and a dial indicator are required. Other manufacturers give a torque reading on the propellor nut. On certain models, Mercury suggests that the gears be coated with Prussian Blue, assembled and rotated, while pushing down on the end of the drive shaft. Then the case is disassembled and the contact pattern inspected.

Obviously the experienced mechanic will not quite follow factory recommendations here. He is working against time and is paid by the work he turns out. His approach is to add slightly more shims then he thinks he will need. Then the unit is assembled and the prop nut turned back and forth. The play in a dry unit can be felt as the gears meet. This bit of feel is the backlash. Gear cases are set up so that there is just a barely detectable bit of play before the teeth touch. On your first few jobs, it might be well to set up the backlash and have an experienced mechanic check it before the engine is started. Most of these men are glad to help a beginner.

OMC ELECTRIC SHIFT

OMC motors (this group includes Evinrude, Johnson, Gale, and some Sea King brands) may be equipped with an electric gear case. Forward and Reverse are selected by a switch which energizes an electromagnet in the gear case. The transmission is mechanically similar to conventional units—it contains a drive gear flanked by forward and reverse pinions—but does not employ a sliding clutch dog to engage the gears. Instead, power is transmitted by coil springs and electromagnets.

Operation

There are two coil springs and two doughnut-shaped magnets, one for each gear. Both gears revolve with the drive shaft at all times. One end of each coil is fastened to the appropriate gear by Allen screws. The free end of the coil rides on the clutch hub. The hub is splined to the propellor shaft. When the gear is not engaged, the coil, rotated by the gear, spins freely on the clutch hub, and no power is transmitted to the propeller. But when one of the electromagnets is energized, the forward (or reverse) coil is attracted to it. The coil stretches along its length and the free end then rides on serrations cut into the clutch hub. These serrations are a friction surface and act to wind the coil around the clutch hub, thus locking the appropriate gear and the propellor into one unit. When the circuit to the electromagnet is broken, the coil retracts to its normal position and rides idly on the smooth surface of the clutch, and no power is transmitted.

The main advantage of this transmission is that shifting can be done by remote control, without the complexity and bother of mechanical linkages. Shifts are quick and silent, without the bump and grind associated with purely mechanical transmissions. The disadvantages are the added parts and potentially troublesome wiring. However, these units are quite reliable in service.

Possible malfunctions are failure to shift into either or both gears, lock up (that is, both gears engaged at once so that the drive shaft cannot move), and failure to disengage from a single gear.

Service

In servicing electro-mechanical devices, it is always wise to check the electrical parts first. This is especially true in marine applications, where the environment is hostile to electrical components. Most electrical faults can be traced to bad or oxidized connections. The hot wire is on the "A" terminal of the ignition switch and goes to the forward and reverse switches. A hot wire from each switch runs to the powerhead, where it is anchored, and down to the lower unit. Forward is color-coded green and Reverse is blue. Each coil has its own ground wire running up to the power head.

You can check the wiring either with a voltmeter or an ohmmeter. Most mechanics prefer the latter since there is no problem of reversed polarities and meter damage. To use the ohmmeter, first disconnect the battery from the system. With the ignition switch "off" there should be zero resistance between the hot side of the switch and the "A" terminal. This means that current will flow through the switch. Next, check the resistance from the "A" terminal to the hot side of the Forward and Reverse switches. Again there should be zero or close to zero resistance. When either switch is depressed, it should pass current. The wiring to the transmission can be checked at the quick disconnect plug on the motor. There should be 7-8 ohms between each hot wire (blue and green) and the powerhead. A reading of zero means that the wiring has shorted to ground, and infinity means that the circuit is open. If the gear box is locked, check the resistance between the green and blue leads. It should be infinity since each is a separate circuit, insulated from the other. Low resistance means that the coils have shorted together, and when either Forward or Reverse is selected, both are energized, freezing the shaft.

Overhaul

To disassemble the gear case, remove the engine cover and free the wires from their anchor lugs. The gear case can be dropped a few inches to reach the disconnect terminals between the electromagnets and the upper wiring. Be extremely careful not to chafe or break the wires. Inspect the water tube inside the shaft housing. If it is clogged, exhaust gasses will burn the wires.

As with any gear case, seemingly identical parts such as pinion gears and clutch hubs should not be interchanged. Special OMC tools are required to remove the drive shaft cup and bearing assembly and the forward coil on some models. On 1966-1968 production, the Reverse gear has forty uncaged rollers between it and the clutch hub.

A very critical part of assembly is the anchoring of the coil springs to the pinion gears. If these springs work loose, the unit will fail. Two sizes of set screws have been used: 8-32 screws should be torqued to 15-20 inch-pounds and 10-32 screws should be tightened to 30-35 inch-pounds. Use Locquic Primer N on the threads and allow the assembly to cure for four hours before handling. The curing process can be speeded by heating the parts in an oven at 250 degrees F. for a half hour.

Three-cylinder Evinrude and Johnson units built since 1968 have another type of electric transmission. These boxes employ a solenoid and oil pump arrangement for shifting. They are somewhat complicated, and require an inventory of special tools to service.

TRANSAXLES

These units combine the functions of gear reduction and differential action. Servicing is similar to transmission work. On some models, there is a difference in length between right and left hand axles and a mistake here might entail a complete teardown. Test the differential by turning the axles in opposite directions. There should be little or no end play and the unit should turn freely.

The "Duo-Trak" (a trade name of the Illinois Tool Works) limited-slip differential is used on some tractors to improve traction in marginal situations. It is not a "locked differential" and if one wheel is jacked up, it will spin. Fig. 6-11 is an exploded view of this device as incorporated in the Peerless series 2300 and 2400 transaxles. Disassembly is by removal of the four through bolts and the snap ring. The brake spring holds the gear pinions in place; this spring can be removed with a large pair of curved snap-ring pliers. To assemble, install the body cores so that the pockets of one core are out of alignment with the pockets of the other core. Install the pinion gears on one side so that the side gear meshes with the five pinions. Turn the unit over and install the pinions on the other side to mesh with the pinions already in place. Insert the brake spring so that it bottoms on the side gear. Most of the ten pinions should be in contact with the spring. Install the axles and torque the bolts to 7-10 foot-pounds.

KICKSTARTERS

Many English and European types work by means of a ratchet which rides on the inner face of a gear. The sear is softer than the gear teeth and wears first. Regrinding or attempts to build up the sear with welding are usually a waste of time. The sear must be replaced (which can mean a long wait on some of the older bikes). Most American and Japanese designers prefer a sector (Fig. 6-8). The first tooth gets most of the wear and often the starter can be repaired by simply grinding that tooth off. But most kickstarter problems are really in the clutch—the numerically high gear ratio of the starter mechanism will accentuate clutch slippage.

CLUTCHES

The controls on small motorcycles, chain saws, reel mowers, and go-karts are often simplified by means of centrifugal clutches. The clutch automatically engages as the engine moves off idle. Designs vary, but all have spring-loaded weights which move friction surfaces into engagement. (Sometimes the weights are the shoes themselves, as in the Mercury clutch.) Repairs are generally no more complicated than cleaning—with detergent and hot water—and periodically replacing or relining the friction surfaces.

Larger motorcycles have traditionally been fitted with multidisc wet plate clutches of the pattern illustrated in Fig.

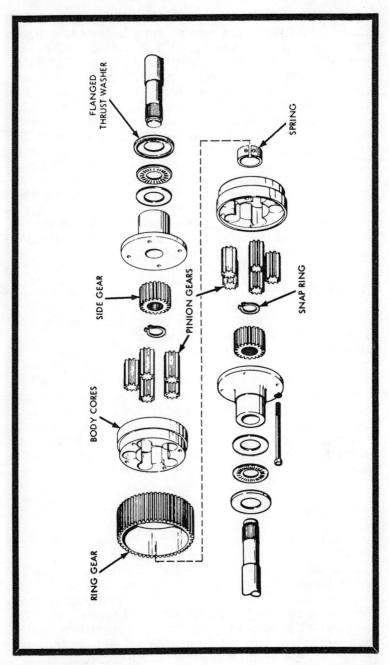

Fig. 6-11. A Limited-Slip Differential used on Garden Tractors and Snow Plows. (Courtesy Tecumseh Products Co.)

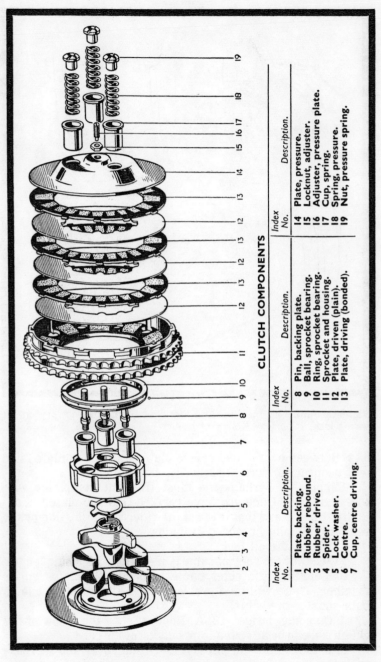

Fig. 6-12. Triumph Clutch with Adjustable Pressure Springs. The screws should be tightened down in small increments.

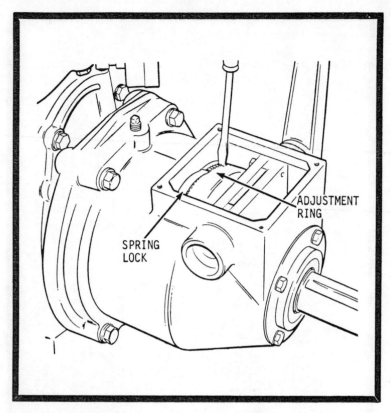

Fig. 6-13. Adjusting the Rockford Wet-Disc Clutch. Do not pry on the spring lock. (Courtesy Kohler of Kohler.)

6-12. Between every steel driven plate is a friction plate lined with neoprene or cork. Oil serves as a coolant.

Slipping is perhaps the most frequent complaint with these clutches. Once you have verified that the clutch is actually slipping and that it is not the rider's imagination or the rear tire slipping, proceed as follows:

1. Check the oil level.
2. See that the control cable is free along its whole length.
3. Lengthen the outer cable to bring the plates closer together. There will be a provision for adjustment at one or both ends of the cable.
4. On some designs (BSA, Hondas, etc.) the throw of the clutch rod can be adjusted from outside of the case.
5. If the clutch still slips, remove the cover and increase the overall spring tension on the plates. Fig. 6-12 shows the typical layout, although this will vary with the manufacturer.

The screws should be tightened evenly and not past the point where the spring coils are collapsed against each other.

6. Finally, the friction discs will have to be replaced.

Most of these clutches "drag," especially when cold. But shifting should be smooth and low gear should engage with only a faint click. Adjust the cable to open the plates, i.e., make the cable effectively longer, and increase the throw of the clutch rod. If the problem persists, tear the unit down and inspect for galling and warped plates.

A number of American industrial engines are equipped with the Rockford toggle clutch. To adjust, disengage the clutch, remove the cover, and turn the adjustment ring one notch at a time (Fig. 6-13). When correctly adjusted, the mechanism will move over center with a firm pressure on the handle.

Dry clutches are found on BMW motorcycles and on the larger industrial engines (Fig. 6-14). On most installations, a single disc is sandwiched between the flywheel and the pressure plate. The disc is lined on both surfaces with an asbestos compound similar to that found on brake shoes. A

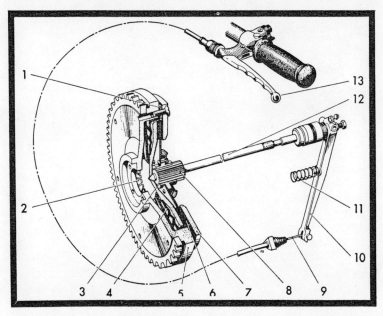

Fig. 6-14. A BMW Single Dry-Disc Clutch. When clutch lever 13 is released, disc 6 is held against pressure ring 7 by diaphragm spring 2. The pressure ring is bolted to flywheel 5. When the clutch is disengaged, push rod 12, working against the diaphragm spring, separates the disc from the pressure ring. (Courtesy of Butler & Smith, Inc., Distributors for BMW.)

throwout bearing may be part of the design. Do not wash these parts in solvent.

Slippage indicates faulty adjustment, oil on the disc from a leaking engine or transmission seal, or lining wear. Failure to disengage may be caused by a defective throwout bearing or by improper adjustment. On disassembly, inspect the flywheel and the pressure plate for scoring.

Chapter 7

Accessories and Controls

Automobile mechanics often find themselves spending a tremendous amount of time on secondary systems such as windshield wiper motors and air conditioning controls. These are essentially "add-ons" and are frequently not as well engineered as the basic vehicle and its power train. The same is true of small engines: some accessories are good, but many are poorly designed and made of inferior materials, and mechanics will spend a great deal of time on them. Sometimes the mechanic will modify the part, or substitute a better one, or simply discard the accessory. The gasoline tank caps with a float and dial arrangement found on some lawnmowers are a case in point. It is an over-refinement to have a fuel gauge on a lawnmower, and this particular item is a fire hazard. Most of these caps leak. The best thing the mechanic can do is to discard the cap and replace it with a stock item.

REWIND STARTERS

Sometimes called recoil starters, these devices are found on outboards, lawnmowers, go-karts, mini-bikes, and on a few of the smaller industrial engines. There are many variations of design, but all of them feature a rope (or a stainless steel cable) which is pulled to start the engine and which automatically retracts when released.

A number of mechanical problems are involved. The starter mechanism must lock solidly to the flywheel when the rope is pulled, and release as soon as the engine starts. A starter which fails to release will be literally exploded by engine torque. Several types of starter clutches have been devised.

Ball and Cam Clutch

Briggs and Stratton uses a ball and cam clutch (Fig. 7-1). Four loose ball bearings are contained in a housing which is affixed to the flywheel. An eccentric cam rides in the housing.

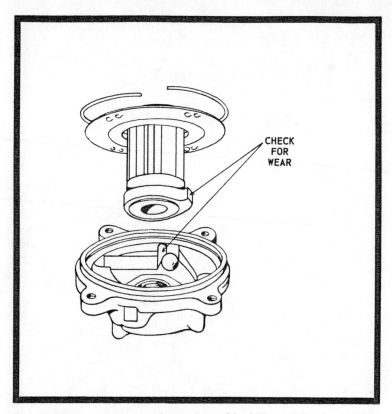

Fig. 7-1. Briggs and Stratton Clutch Assembly. Late models may be pried open with a screwdriver blade. (Courtesy Briggs and Stratton.)

As the rope is pulled, the cam turns and traps one of the balls against the housing. Once the engine catches, the housing rotates relative to the cam and releases the ball. These clutches are almost foolproof and will give good service if they are kept dry. It is a mistake to oil the ball bearings, since the oil will attract dirt which will cause the bearings to stick. However, some lubrication may be necessary on the upper portion of the crankshaft. Use a very small amount of high temperature lubricant such as wheel bearing grease or breaker point lube. These clutches may be removed from the flywheel as an assembly by tapping the lugs with a soft mallet. (Briggs supplies a special spanner wrench for this job.) Standard right-hand threads are used. To asemble the unit, place the cam in the housing and install the bearings around the cam.

Lawnboy uses a spring-loaded sear on the flywheel (Fig. 7-2). When the engine is not running, the sear locks into the starter pulley. As the pulley turns, the sear transmits this motion to the flywheel. Upon starting, centrifugal force moves the sear out of engagement. These units must be kept clean and free of lubricant. Occasionally a sear spring will break.

Ratchets

Most other starters, Clinton, Fairbanks-Morse, Tecumseh, etc., depend upon a ratchet. Generally there is a friction brake to transmit energy to the ratchet (much like the friction spring on a typical bicycle coaster brake). Look at Fig. 7-3. The ratchets (friction shoe plates) are shown at No. 11 and the brake spring is No. 15. A number of problems can occur with this design: the brake spring can become worn beyond acceptable limits and refuse to activate the ratchets; on some, the ratchets can get dull in service and slip against the hub; and the mounting screw can work loose, relaxing tension on

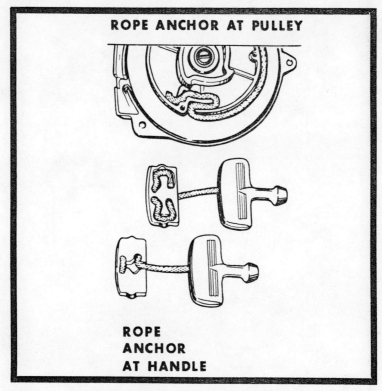

Fig. 7-2. Rope Attachment on "Lawnboy" Starter. On most other designs, the rope is secured with a knot and heated. (Courtesy Outboard Marine Corp.)

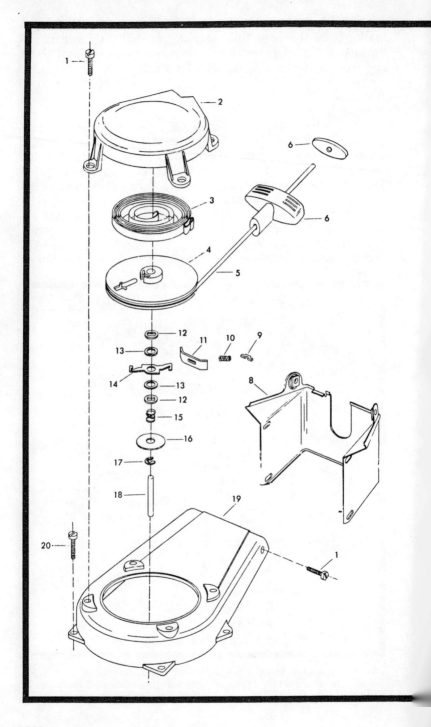

ILLUS. NO.	PART NO.	QTY.	DESCRIPTION
1	1096	6	Pan hd. screw w/LW, 1/4 - 20 x 1/2
2	15546	1	Cover
3	15528	1	Rewind spring
4	15522	1	Rotor
5	15525	1	Cord
6	15540	1	T-handle w/insert
8	236648	1	Cylinder cover
9	15518	2	Spring retainer plate
10	15503	2	Friction shoe spring
11	15519	2	Friction shoe plate
12	15510	2	Brake washer
13	15532	2	Fiber washer
14	15501	1	Brake lever
15	15504	1	Brake spring
16	15511	1	Brake retainer washer
17	15513	1	Retainer ring
18	15508	1	Centering pin
19	175596-1	1	Fan housing
20	1282	4	Pan hd. screw, w/LW, 1/4 - 20 x 5/8
	A27395-4	1	Identification plate w/screws (not shown)
	15535	1	Friction shoe assembly, includes items 9, 10, 11 and 14 (not shown)
	175063-1	1	Complete starter (not shown)

Fig. 7-3. Rewind Starter fitted to West Bend Series "L" Engine. (Courtesy Chrylser Corp., Outboard Div.)

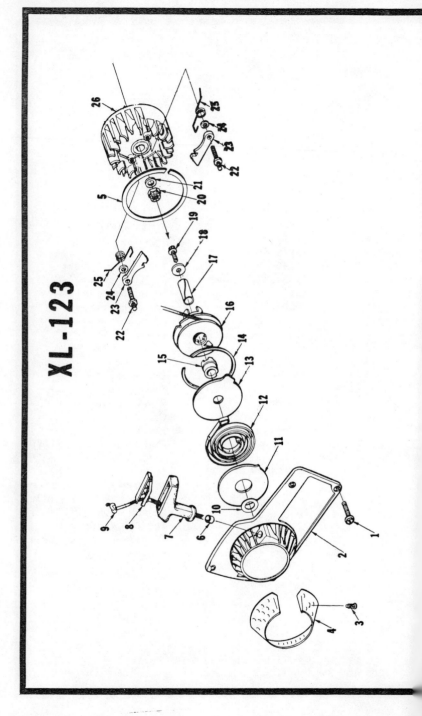

No.	Description	Part No.	Qty.
1	SCREW-hex, 8-32 x 1/2	80983	4
2	HOUSING-fan	A-67192-A	1
3	Includes: STUD-drive	67191	4
4	SCREEN-fan housing	65491	1
5	RING-air flow	67222	1
6	BUSHING-starter rope	67177	1
7	GRIP-starting rope	64155-A	1
8	INSERT-rope retaining	64154-A	1
9	ROPE-starter	63869	1
10	WASHER	67185	1
11	SHIELD-spring (inner)	63713	1
12	SPRING-rewind	63870	1
13	SHIELD-spring (outer)	63714	1
14	RING-shield retaining	63875	1
15	LOCK-spring	64948	1
16	PULLEY & CUP ASSEMBLY	A-63952-B	1
17	BUSHING-starter post	64374	1
18	WASHER-retaining	63606	1
19	SCREW-hex, 10-32 x 1/2	80648	1
20	NUT-lock, 5/16-24	81112	1
21	WASHER-flat, 5/16	84065-1	1
22	STUD-shoulder	63877	2
23	PAWL-starter	64442	2
24	WASHER-flat	58851	2
25	SPRING-starter pawl	58758-2	2
26	ROTOR-magneto	63637	1

Fig. 7-4. Starter for Homelite XL-123 Chain Saw. (Courtesy Homelite, A Textron Division.)

the whole assembly. The latter is fairly common, and a permanent repair can be made by cleaning the threads and coating them with LocTite. Worn brake springs should be replaced, and the ratchets can be sharpened. On some models it is possible for the serrations on the flywheel hub to strip. When replacing the hub be sure that the correct part is used, since there have been a number of variations of the height of this part. Another peculiarity of many ratchet starters is that they can be assembled wrong. If the ratchets are upside down, the clutch will slip. This has not been done to confuse mechanics—these starters are designed to be used on either counter or clockwise rotation engines. The ratchets, main spring, and rope lay can be reversed.

Rope-Wind Springs

Besides the clutch, a rewind starter must have a spring to wind the rope. With the exception of Chrysler-West Bend outboards, these springs are flat coils. The West Bend uses a vertically mounted "rat-trap" spring which should not be disassembled without special factory tools. Springs are installed with a pre-load, so that all the slack on the rope will be taken up. The amount of pre-load varies with the design and with the amount of tension in the spring. A weak spring will have to be pre-loaded more than a new one.

On larger diameter units, such as found on outboards, one turn after the rope is fully wound is enough. Smaller engines may call for six to eight turns. Use no more tension than required to smartly retract the rope because the tighter the spring is wound, the sooner it will fail.

Two methods are used to pre-load the spring. Some starters have a notch cut in the side of the pulley. Fig. 7-4 shows a Homelite starter. The notched pulley is clearly visible at No. 16. A Tecumseh design is shown in Fig. 7-5. Other starters do not have this feature, and the rope is simply wound around the pulley and "fished out" with a pointed instrument. (Some mechanics modify these starters by grinding a notch on one side of the pulley. If you do this, be certain that there are no sharp edges to damage the rope.) The pulley is held with thumb pressure, and the handle is installed.

It is wise to wear safety glasses when handling these springs. The spring itself (with the exception of the Chrysler-West Bend) is relatively weak. But it can damage the eye. It is also good practice to release tension from the spring prior to removing it. On Briggs engines, the spring protrudes from the housing. This free end should be pulled as far as possible out of the housing with a pair of pliers. On other designs, the handle is removed from the rope and, using the thumbs as a brake, the pulley is allowed to revolve and release the spring tension (Fig. 7-5).

Springs can fail in a number of ways. On Briggs engines it is not unusual for the spring to be "swallowed" by the housing. The free end slips out of its position. Generally, repairs can be made by carefully forming the lug on the housing. In extreme cases, the housing will have to be replaced. Any of these springs can fatigue and break. The break will occur at the fixed end or at the pulley. Some mechanics attempt to repair the spring by grinding new ends on it. Generally this practice is a waste of time, although it can result in a short term repair where new parts are not available. In any case, do not heat the spring to bend it. Weak springs can be corrected by increasing the pre-load, although there are limits to this.

Unlike most other recoil starters, the Briggs design employs a spring which must be wound on the pulley (Fig. 7-6A). The spring is hooked to the pulley and extended through the hole in the starter housing. The inset in the drawing shows the proper clearance between the nylon bumper and the pulley.

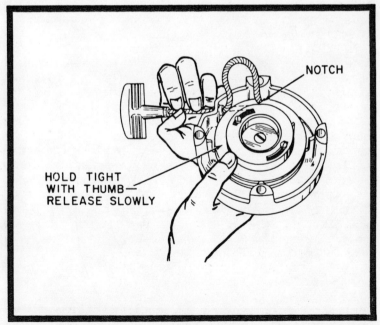

Fig. 7-5. Removing Spring Tension on Starter with Notched Pulley. (Courtesy Tecumseh Products Co.)

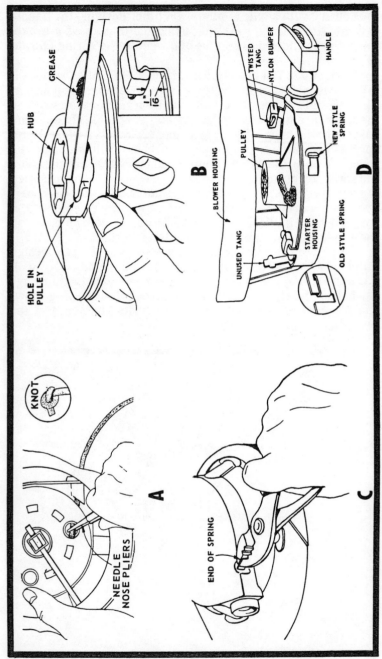

Fig. 7-6. Installing and Pre-Loading Spring in Pull-Type Starter. (Courtesy Briggs and Stratton.)

Using a piece of ¾" stock and a wrench, wind the pulley 13¼ turns counterclockwise (Fig. 7-6B). As the pulley is wound, the spring will move into the housing and snap into place. Holding tension on the pulley, insert the rope (with the handle secured) through the rope guide and out the hole on the pulley. (On old style starters you will see a guide lug on the inner surface of the pulley. The rope must pass inside, i.e., toward the center of the pulley, of this lug.) Once the rope is knotted, tension on the pulley is released and the rope is retracted.

To remove spring tension from most Briggs starters, pull the spring out of the housing as far as possible (Fig. 7-6C). Fig. 7-6D shows a Briggs & Stratton starter in inverted position. You can see how the spring tang, extending through the starter housing, simplifies servicing.

On all other engines, the spring is entirely contained in the housing. New springs are supplied already coiled in a ring of heavy metal. The spring is positioned in the housing, with the loop directly over the post, and pushed out of the holder. Mechanics save one of these holders in case they have to rewind a used spring. It is much easier to wind it in the holder than on the engine.

Once the spring is in place, the pulley can be positioned over it. Generally there will be a notch in the pulley which connects to the spring. Oil the assembly lightly.

So far we have been discussing American designs. The Japanese Honda industrial engine has a starter similar to the American types, and should not give any difficulty to someone who has worked on Clinton, McCulloch, or Tecumseh engines. However, the German JLO engine is an entirely different proposition. These engines are widely used on go-karts and come in both left and right hand rotation. Before disassembling, obtain three metric studs which are placed in holes around the flywheel housing. These studs hold the spring in place. Likewise, the studs must be in position when the spring is installed.

Rope and Cable

Most rewind starters use nylon rope. It can be purchased as a pre-cut item for each engine, or it may be purchased in bulk. Be sure that the diameter and the length of the replacement rope is the same as the original. On older engines, the rope is knotted at the pulley and at the handle. Newer designs use a friction lock. Fig. 7-2 shows a typical arrangement as used on Lawnboy mowers and on many

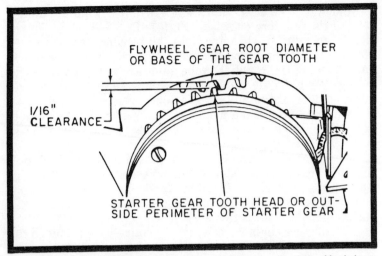

Fig. 7-7. Vertical-Pull Starters must be positioned so that the Mesh is as shown. Too deep a tooth contact will destroy the starter. (Courtesy Tecumseh Products Co.)

outboards. Always melt the ends of the rope by heating or by dipping in a solvent such as acetone. This will prevent fraying. Nylon ropes usually break at the pulley end. Of course, they can be shortened and re-tied, but this is not advisable. Too much oil in the starter will cause the rope to swell and stick.

Stainless steel cables are sometimes encountered. If the cable shows any signs of fraying, it should be replaced. A single broken strand can cause jamming.

Mount the starter on the flywheel shroud. Turn the engine a few times before starting. If the starter drags, loosen the bolts and reposition it.

Vertical-Pull Starters

Both Briggs and Stratton, and Tecumseh, supply vertical-pull starters for rotary lawnmower applications. These starters make life a little easier for the user, since he does not have to bend over the engine. The main difference between these and other rewind starters is the gear which engages teeth on the rim of the flywheel. The mesh between this gear and the flywheel is critical. Too little contact and the starter will not engage, too much and it will not disengage (Fig. 7-7). Generally, service is the same as for conventional units. The Briggs design is quite similar to Tecumseh starters. Most manufacturers suggest that the engagement drive should not be oiled.

Spring and slack adjustments for a Briggs and Stratton starter are shown in Fig. 7-8. In A, slack is provided as shown. As with all rewind starters, there must be no tension on the rope when the pulley is removed. In Fig. 7-8B, after the spring tension is released, the plastic cover may be pried off, and the spring anchor removed. On assembly (Fig. 7-8C) retract the rope by turning the pulley counterclockwise. Snap cover in place, pull the rope off the pulley for about one foot (D), and wind the pulley two or three turns counterclockwise to give the proper rope tension. All that remains is to install the starter with proper gear mesh on the flywheel.

A Tecumseh vertical pull unit is shown in Fig. 7-9. Repair procedure on this starter is as previously outlined.

Some customers seem to be particularly hard on rewind starters. It is not usual for the same customer to come back week after week with broken ropes and broken springs. Have the customer attempt to start the engine. Generally, he will jerk on the rope and keep pulling on it past the end of travel. It takes tact, but the customer should be shown how to use the starter properly. First, take up all tension in the mechanism by pulling on the rope for a half inch or so. Then with a smooth, fluid motion, pull the handle through to a point about three quarters the length of travel. Hold the handle as it retracts; never allow the handle to snap back.

Rewind starters can also fail because of overuse. Their designed life-span is well beyond that of the engine, but a hard-to-start engine will cause wear beyond design limits. Sometimes the cure to a starter problem is a good tune-up.

IMPULSE STARTERS

Also known as crank, ratchet, or wind-up starters, these units are found on lawnmowers, snow plows, and cultivators. A heavy spring provides energy to spin the engine. It is wound by a crank and ratchet. Fifty years ago, a few motorcycles had a similar arrangement, except that the engine did the winding. All the operator had to do was turn a lever and the spring released. Today, the operator must do the cranking himself.

The major advantage of an impulse starter is convenience: it is physically easier to wind a ratchet than to pull on a starter cord. These starters are especially popular with elderly people. Another advantage is that the starter gives a tremendous kick to the engine, and should provide faster starts.

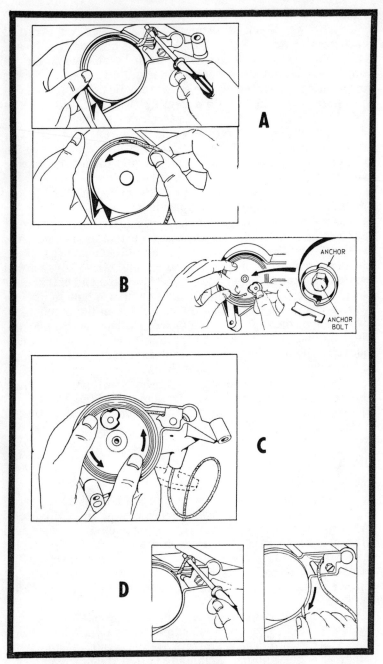

Fig. 7-8. Briggs and Stratton Vertical-Pull Starter Slack and Spring Adjustment. (Courtesy Briggs and Stratton.)

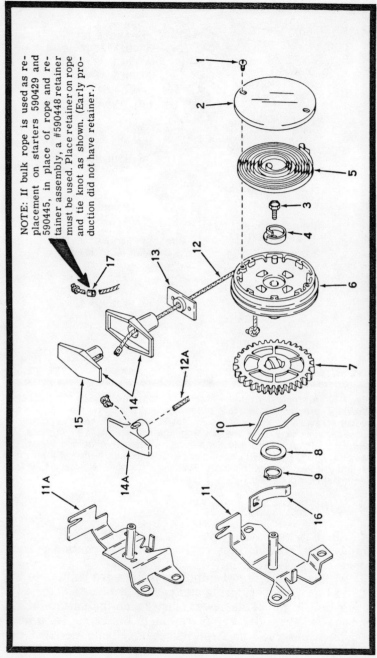

Fig. 7-9. Tecumseh Vertical-Pull Starter. (Courtesy Tecumseh Products Co.)

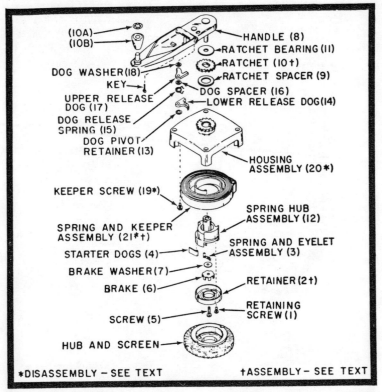

Fig. 7-10. A Typical Impulse Starter. Disassembly is in the sequence shown. The asterisks indicate potential danger points. No. 19—remove keeper screws only when necessary; No. 20 remove the spring and keeper by pounding one foot of the housing against a hard, flat surface. Be certain that the spring and keeper have fallen free before lifting the housing; No. 21 replace the spring and keeper as a unit—do **not** attempt to rewind the spring on this unit. Assembly is the reverse. (Courtesy Tecumseh Products Co.)

Safety Precautions

All that we have previously said about safety applies doubly here. The springs on these starters contain enough energy to break an arm.

At this writing, most impulse starters are similar to the one in Fig. 7-10. The spring and casing must be treated as a single unit. Some of the newer starters, on the pattern of the one shown in Fig. 7-11, can be rewound. But do not try it until you have observed an experienced mechanic perform the operation.

Occasionally you will get a starter with a cracked housing. I have seen mechanics turn pale when a customer walks into the shop with one of these "grenades" and slaps it on the counter. If the unit is still on the machine, climb up on a bench and remove the mounting bolts with a long extension bar. Once removed, the housing should be put in a trash barrel and broken up with a heavy weight.

A frozen engine is another problem. The customer will have wound the starter, released the trigger, and brought it to you. At this point it is tempting to refuse the job. But somebody has to disarm that starter. Try to free the engine by striking the blade with a two-by-four. If this doesn't work, stand on the bench and, again using a long extension, loosen the mounting bolts one at a time. Carefully loosen the next to the last bolt. The starter will release, pivoting on the remaining bolt. If the bolt pulls loose, the starter will bounce off the ceiling.

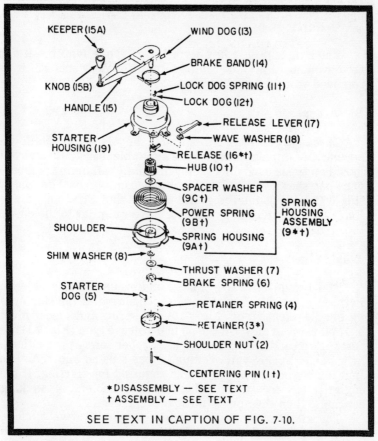

Fig. 7-11. Newer Type Impulse Starter. (Courtesy Tecumseh Products Co.)

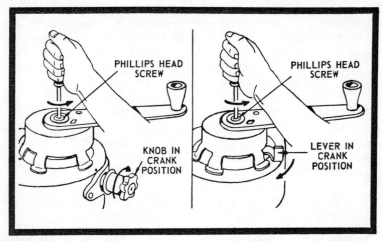

Fig. 7-12. Releasing Spring Tension on Impulse Starter. (Courtesy Briggs and Stratton.)

The Briggs design has a very worthwhile feature—it can be unwound (but not completely disarmed) before it is removed from the engine (Fig. 7-12).

Basic Types

Three basic types, in various models with detail differences, of impulse starters are available. Figs. 7-10 and 7-11 show two types used on Tecumseh (and Craftsman) engines. (The one shown in Fig. 7-10 is now becoming obsolete, but is still widely encountered.) The Briggs and Stratton unit is shown in Fig. 7-12. Pincore uses a starter similar to the one in Fig. 7-10.

Trouble Spots

The Tecumseh starters have had a number of problems. Perhaps the most frequent is a broken spring. If the starter is the handle release type (Fig. 7-10), the spring and keeper must be replaced as a unit. On the trigger release type (Fig. 7-11), a new spring can be installed but use proper safety precautions. Any starter spring will eventually lose tension and refuse to spin the engine the 90 rpm or so required for starting. If the engine turns freely, and there is no problem of sticking bearings, then the spring must be replaced.

The clutches on these starters may also give trouble. Expect considerable wear on the hub and screen (shown at the bottom of Fig. 7-10) and on the starter dogs. The dogs may be sharpened as indicated in the section on rewind starters. The brake spring should be replaced as a routine matter when the unit is disassembled. Wear on the spring will prevent the dogs from extending out of the retainer. The screw in Fig. 7-10 must be tight, as must the shoulder nut in Fig. 7-11. Looseness here will prevent the dogs from engaging.

If the starter releases before it should, examine the upper and lower release dogs (No. 17 and No. 14 in Fig. 7-10) or the release lever (No. 17 in Fig. 7-11). This lever pivots on a rivet which can work loose. If that is the case, grind the head of the rivet flush with the housing, being careful not to overheat the wave washer (No. 18 in Fig. 7-11), and punch it out. Replace with a quarter inch bolt and lock nut. Inspect all springs for breakage. Lubricate with heavy oil.

Clinton engines are equipped with two types of rewind starters which are very similar to the Tecumseh units discussed in the preceding paragraphs. The early Clinton starter is unusual in that it features a gear reduction on the crank. When installing the large gear, the beveled side of the teeth must be toward the open side of the housing. Clinton suggests that their starters be coated with liberal amounts of high temperature grease such as Lubriplate.

All Briggs units are similar to the type shown in Fig. 7-12. The early production type has a knob which is rotated to hold the flywheel against spring tension. Damage will result if the knob is turned to "Start" while the engine is running. Later models have a lever which engages teeth on the starter spring housing. Some Briggs starters have the additional refinement of a ratchet which allows the crank to be reversed when there is not room for a full turn of the crank. These starters are usually found on horizontal crankshaft models.

Spring tension can be relieved by loosening the screw on top of the assembly, with the control in the crank position. BUT THE SPRING ITSELF MUST NOT BE REMOVED FROM ITS STEEL HOUSING. TO DO SO IS DANGEROUS. On late models, the housing teeth can wear and cause the starter to slip as it is being wound. Late models also have a spring washer which is placed between the top of the spring housing and the blower shroud. This washer must be replaced if it is broken.

REMOTE CONTROLS

Following a time-honored practice, motorcycles employ flex cables to operate the throttle, clutch, and front brake.

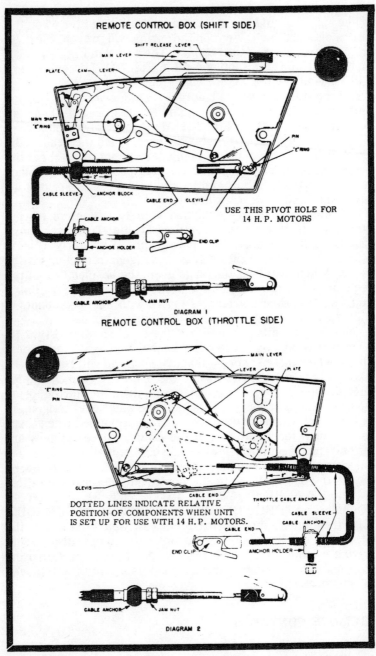

Fig. 7-13. Remote Control Box used on various "Scott" Outboard Engines. (Courtesy McCulloch Corp.)

Some bikes use cable on the rear brake as well, although there is a feeling among riders that the rod arrangement is more sensitive. The outer casing is anchored at both ends and threaded for adjustment. The inner cable is made of fine wire for flexibility and usually has a brass ball end soldered in place. If you replace one of these ends, use plenty of heat and acid—core solder. Outboard throttle and shift controls use a similar cable. These cables should be kept lubricated and periodically inspected for kinks and broken strands. A single broken strand, doubled up on itself, can cause the cable to stick.

Remote control boxes for outboards can be quite complex (see Fig. 7-13) and should be mounted with care. It is a good idea to have the customer "try on" the control before it is finally bolted up. Motorcycle control levers should be the ball end type for safety, and unless the rider is very experienced, discourage the use of one-eighth turn twistgrips.

Bowden cables are used on lawnmowers and industrial applications. The outer casing is similar to motorcycle casing, and one end is adjustable by means of a friction clip. The inner cable is a single strand of spring wire, hooked on one end and usually secured by a set screw on the other. Lubricate with a small brush dipped in oil and inspect the inner wire for bowing as it leaves its casing.

Rather than use a factory replacement, it is cheaper to purchase Bowden cable in rolls and cut to length. Using a three-cornered file, cut the casing most of the way through. Do not go too deeply because a nick on the inner wire will ruin it. Hold the cable with one hand on either side of the cut and twist in the direction of the lay. The cable will stretch and snap. The inner wire can be snipped with a pair of large side cutters or lineman's pliers. Bend a hook on the wire to match the old one. Try to get it right the first time; these spring steel wires are brittle.

Secure the cable at the control end—most Bowden cables screw into the control lever—and adjust the free end so that it fully opens and closes the throttle plate. On engines with integrated controls working from a single cable, adjustments are a bit more critical. The kill switch, wired to the magneto primary side, shorts the ignition when the throttle is moved past idle (Fig. 7-14). The choke engages at the other extreme of movement, past the full throttle position. The choke must open fully when the handle is in the "Choke" or "Start" position, but must not be open in the high speed or "Run" position. Often cable adjustments are not enough, and it is necessary to bend the linkages (Fig. 7-14C).

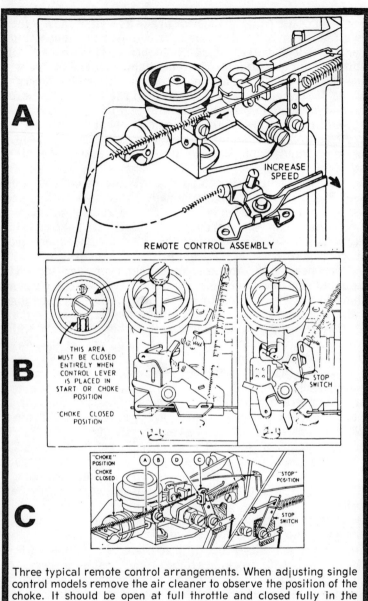

Three typical remote control arrangements. When adjusting single control models remove the air cleaner to observe the position of the choke. It should be open at full throttle and closed fully in the CHOKE or START position. Usually it is enough to adjust the Bowden cable at B although it may be necessary to bend the link or actuating lever. When the control lever is moved past idle, it should contact the stop switch and short the points.

Fig. 7-14. Three Typical Remote Control Units. (Courtesy Briggs and Stratton.)

GOVERNORS

All industrial engines are governed to limit the top rpm and, in most cases, to provide compensation for sudden loads. Every manufacturer seems to have his own favorite way of achieving this, and adjustment procedures vary. Most mechanics set the governor by ear, but a tachometer should be used. Electric tachometers specifically designed to handle the high primary current developed in magneto engines are useful, but quite expensive. On engines without external "kill" switches, the flywheel must be removed to make the connections for electric tachs. A better solution is the vibrating reed type meter which is held in the exhaust and works on the principle of a tuning fork.

Air Vane Type

Air vane governors are common on the smaller engines. A hinged vane is mounted in the air stream created by the flywheel fan (Fig. 7-15). The vane is connected directly or by

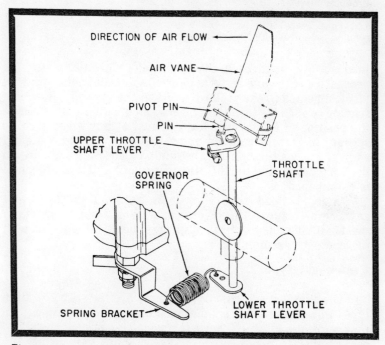

Fig. 7-15. An air vane governor of the type used on Power Products two-cycle engines. Other designs use a linkage between the vane and the throttle plate. (Courtesy Tecumseh Products Co.)

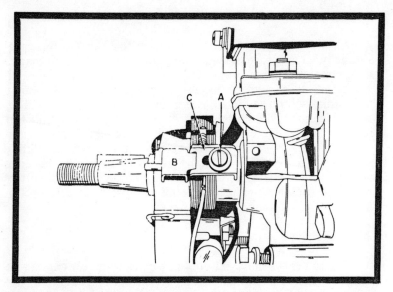

Fig. 7-16. A Speed-Limiting Centrifugal Governor as used on Power Products Engines. To increase speed, loosen "A" and move the bracket toward the flywheel. These engines are factory set at between 4,400 and 5,000 rpm. Do not attempt to make this adjustment while the engine is running. (Courtesy Tecumseh Products Co.)

means of a linkage to the carburetor throttle plate. Air pressure on the vane tends to close the throttle, but a spring is also connected to the plate. It pulls in the opposite direction, tending to open the throttle. The amount of pull is controlled by the position of the throttle lever. Thus, at any given throttle setting, the forces balance and the engine runs at a steady rpm. If the load increases, the engine slows, air pressure on the vane drops, and the throttle opens until a new equilibrium is established.

Under no-load conditions it is not unusual for these governors to surge; however, carburetor adjustment cures most of it. Stretched or broken springs should be replaced with the correct part number.

Centrifugal Type

The simplest centrifugal governors are the speed limiting type. Power Products engines are sometimes fitted with a magneto cut-out, which shorts the points at a pre-set speed. Do not attempt to make adjustments on these units while the engine is running (Fig. 7-16).

A chain saw governor is shown in Fig. 7-17.

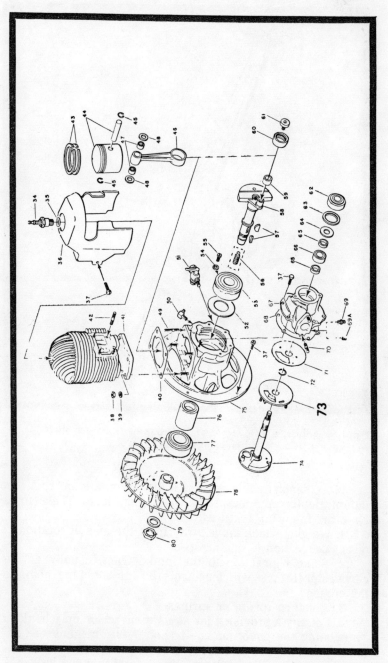

Fig. 7-17. Exploded Drawing showing the Homelite Series 9 Engine. The governor is shown at No. 73. (Courtesy Homelite, A Textron Division.)

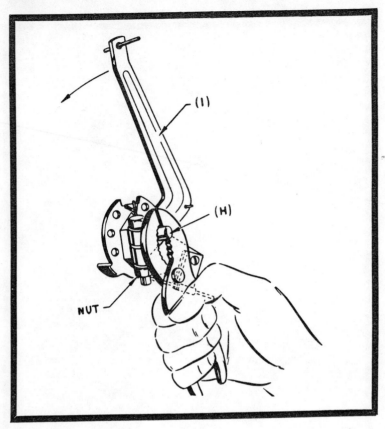

Fig. 7-18. Adjusting a Governor by Twisting the Governor Shaft, Kohler K91 Engine. (Courtesy Kohler of Kohler.)

More sophisticated designs limit the maximum rpm and automatically compensate for changing load. These units have the following features:
 1. Weights which move away from the axis of rotation in direct proportion to engine speed.
 2. Some connection to the intake tract, usually at the carburetor throttle, which controls the amount of fuel going to the engine.
 3. A return spring or springs.
 4. Usually a provision for adjustment which may be done by twisting the governor shaft (Fig. 7-18); bending the actuating link (Fig. 7-19); or by changing the mechanical advantage of the mechanism by using alternate mounting holes for links and springs.

These governors are adjusted so that the throttle is wide open with the engine not running. Check maximum rpm with a tachometer. In Fig. 7-18, the nut which holds the governor shaft H to the governor arm I is loosened. The shaft is turned counterclockwise; the arm is pulled all the way to the left (away from the carburetor). The process is similar on the Tecumseh four-cycle. In Fig. 7-19, the governed speed on some Lawn-Boy engines is changed by bending the linkage. This engine may be equipped with a 3200 rpm governor return spring or a 2800 rpm spring.

It is rare for any governed engine to have a no-load speed in excess of 3600 rpm. Rotary lawnmowers are especially critical because blade tip speed should not exceed 19,000 feet per minute. The following chart (Courtesy of Briggs and Stratton) shows the absolute top speed. The engine should be governed at least 200 rpm **below** this maximum.

BLADE LENGTH (IN.)	MAXIMUM ROTATIONAL RPM
18	4032
19	3820
20	3629
21	3456
22	3299
23	3155
24	3024
25	2903
26	2791

COMPRESSION RELEASES

A few imported motorcycles are still equipped with compression releases to make starting a little easier and to provide some compression braking. Fig. 7-20 illustrates the typical poppet valve type as used on a small motorcycle. During overhauls, the valve should be lightly lapped with grinding compound.

Briggs and Stratton, that most innovative firm, has developed an automatic compression release which raises the exhaust valve off its seat during cranking. The engine still develops enough compression to start, and once running, the valve closes completely.

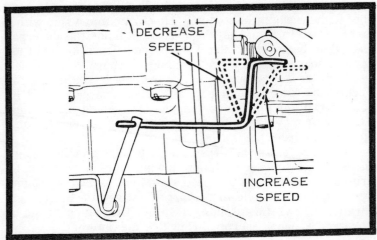

Fig. 7-19. Adjusting a Governor by Bending the Actuating Link. (Courtesy Gale Products.)

Sometimes you will have a customer who complains that he is not physically able to crank the engine. There are add-on electric starter kits, but these are quite expensive. A partial solution is to use two cylinder head gaskets. Another solution, and one which I cannot entirely vouch for, is to drill a ⅛" hole

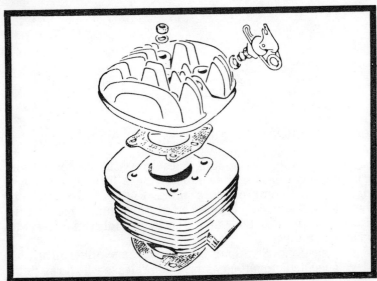

Fig. 7-20. Typical Compression Release Poppet Valve. Note: some owners put another spark plug in this hole and parallel it with the other plug.

in the floor of the chamber about a half inch from the exhaust valve seat. The hole is angled to empty in the exhaust port.

PUMPS

In addition to fuel and oil pumps, small engines may be equipped with pumps for cooling and lubrication of accessories. Except for a few air-cooled models and the most primitive water-cooled types, outboards have a water pump built into the lower unit. The pump may include a bailer. Check the pump by observing the water flow from the outlet, or less accurately, by means of engine heat. Run the motor for a few minutes and stop it. Hold your hand on the water jacket. If the pump was working you will feel a build-up of heat, showing that coolant had been circulating.

Most outboard pumps operate on the same principle as the Scott unit shown in Fig. 7-21. An off-center shaft causes the vanes on the impeller to flex as it sweeps the pump cavity. At low speeds it has the positive action of a displacement pump, and at high speeds it has the efficiency of a centrifugal pump. The unit pictured in Fig. 7-22 shows a second impeller which acts as a bailer, drawing water from the bilges and expelling it through the exhaust system. Small holes in the engine pump casing bleed water to lubricate the second impeller when the bailer is not in use.

Although new materials have improved impeller life, this part can still be a problem. The vanes tend to lose their elasticity and, at least on some engines, stick to the housing in

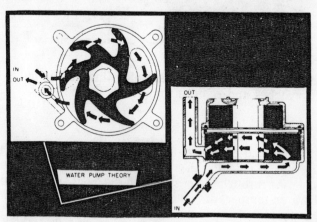

Fig. 7-21. Operation of Outboard Engine Water Pump. Some designs may have an elliptical casing rather than the off-center shaft. (Courtesy McCulloch Corp.)

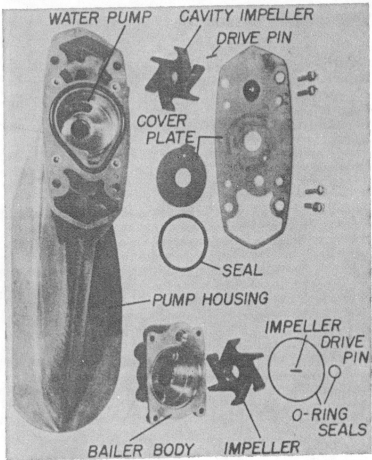

Fig. 7-22. Typical Water Pump-Bailer Unit for an Outboard. (Courtesy McCulloch Corp.)

storage. When replacing the impeller, inspect the pump cover for scratches and replace the seals.

THERMOSTATS

Most liquid-cooled engines use a thermostat to give faster warm-ups. Typically the thermostat prevents water from circulating through the passages in the head (Fig. 7-23). When the water temperature increases past a set point, the stat opens. In industrial engines, the opening temperature varies from 160 degrees to 190 degrees F. Generally a high temperature thermostat means that the engine will run more efficiently. However, if salt water is used as the cooling

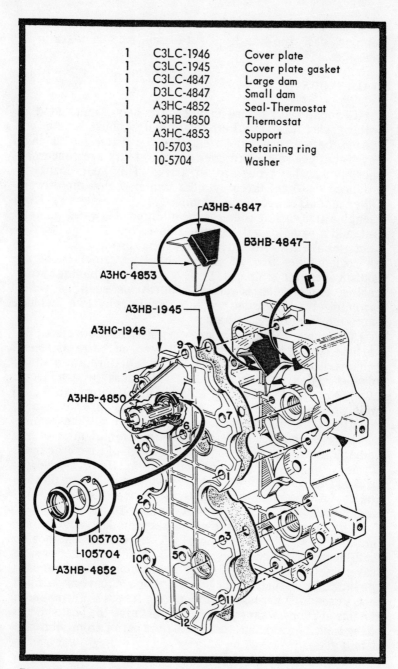

Fig. 7-23. Thermostat Replacement Kit for 60 HP Scott Outboard. (Courtesy McCulloch Corp.)

medium, most mechanics prefer a 150 degrees stat. At higher temperatures, the salt drops out of solution and cakes in the passages.

RADIATORS

Gradually, air-cooled engines are being supplanted by water-cooled designs. Perhaps the air-cooled single will always be with us—who would want a water-cooled chain saw?—but where performance and durability are concerned, the liquid-cooled types are attractive. This year Suzuki introduced a 750 cc, three cylinder, water-cooled motorcycle. Other manufacturers have promised similar machines. Even though water cooling requires the addition of a water pump, a fan, and a radiator with assorted hoses and fittings, the advantages are attractive. These machines are quieter than the air-cooled types and, of course, have extremely accurate control of engine temperatues. It is not impossible that water cooling will spread to other applications such as riding lawnmowers and garden tractors. That's why I've included some information on radiators.

Two types of radiators are currently in production. The earliest type, and one that is still seen on large industrial engines, is the tubular fin type. These radiators are made of rows of vertical or horizontal tubes which pass through fins set at right angles to the tubes. Tubular fin radiators were used widely on pre-World War II automobiles. They are strong and, once the tank is removed, can be "rodded out" to remove corrosion. Cooling is a function of the frontal area of the radiator and of the number of fins per inch of tube. The more fin area, the greater the heat dissipation. However, closely packed fins will require a stronger fan to pull the air through the radiator, and therefore some power is lost which might otherwise be used to turn the crankshaft.

The tube and center type is used on modern automobiles and may well be the radiator which will find acceptance on small engines. Its great advantage is that it cools better with less weight. The fins are arranged accordion-fashion between the tubes. Each fin has a half dozen or more louvers stamped in it to send air across and around the tubes. Unfortunately, the fins are somewhat fragile and do not give the best support to the tubes. As a result, these radiators suffer from vibration damage.

Modern radiators of either type are always fitted with a pressure cap. This is to increase the pressure in the radiator to some figure above sea level. Most caps are set to hold seven to fifteen pounds of pressure. The pressure alone does not help the radiator cool, but it does prevent the water from boiling. For each pound of pressure, the boiling point is raised three degrees. Thus, a ten pound cap would mean that the engine could operate indefinitely at 232 degrees F.

Another way to increase the boiling point is to add ethylene glycol (Prestone or the equivalent) to the water. A 50-50 solution raises the boiling point to 225 degrees. Add a ten pound cap and the boiling point jumps to 255 degrees.

These high temperatures are healthy as long as water loss can be controlled. In the automobile industry, engine temperature has risen with every model year. Today, most engines operate at 195 degrees. This means less wear on mechanical parts, cleaner oil (because sludge formation is reduced), and longer oil change intervals.

Radiators are subject to corrosion build-up and vibration damage. The latter can be minimized by isolating the radiator with rubber shock mounts. Corrosion can be controlled by draining and flushing at least once a year and by adding a rust inhibitor. Most permanent type anti-freezes contain some rust inhibitors, but the "pros" say that this may not be enough.

Minor corrosion build-up can be cured by using one of the chemical cleaners, such as marketed by DuPont. Be sure the solution is thoroughly drained, at the radiator and the block, and neutralized. A surer method is to remove the radiator and have it boiled out at a shop specializing in this work. In some cases, the tank may have to be removed and the water passages cleaned individually. The shop can also test the radiator for pressure.

Small leaks can be soldered. Clean the surface with a wire brush and sandpaper to remove all scale. Then daub muriatic acid on the area to get it chemically clean. Be thorough, since a film of corrosion a molecule thick will prevent the solder from adhering. Use acid-core solder and a heavy iron. If the solder bubbles, the surface is still dirty. Some mechanics solder with a propane torch, but this is difficult. The corrosion problem is increased, and, if you are not careful, the heat from the torch will loosen other joints.

If the radiator loses water through the overflow, inspect the seal around the pressure cap. The cap itself can be tested for opening pressure by a radiator shop or by one of the better equipped service stations.

OIL COOLERS

Oil coolers—really no more than tubular fin radiators—are becoming popular on high performance four-cycles. BSA and Triumph employ them on their 750 cc road machines, and they have been fitted to racers for years. Specialty builders have added oil coolers to a wide variety of engines. They must be connected in series with the oil pump so that all the oil passes through the cooler. A late model Volkswagen oil cooler is a favorite for this kind of adaptation. The early models were soldered, and can fail in extreme high temperature operation. Another choice is the Cadillac power steering oil cooler. It is more compact than the VW and can be purchased quite reasonably in wrecking yards.

Oil coolers should be tested by plugging one end and immersing the unit in water. High pressure air (at double oil pump pressure) is applied to the open end. Brass types can be easily soldered, once the device is thoroughly cleaned to remove all traces of oil.

MUFFLERS

Small engine mufflers are usually constructed of pressed steel with a series of internal baffle plates. In some cases, the muffler may be partially filled with steel shavings or fiberglass mat. Rust, accelerated by the acid and water in the exhaust, is the great enemy. Two-cycle mufflers usually can be taken apart for carbon removal. A faster way—with steel mufflers—is to soak the whole assembly in a strong solution of caustic soda and water. In a few hours all the carbon and rust will dissolve.

In passing, it must be remarked that unmuffled engines are not necessarily more powerful. On two-cycles, removing or "gutting" the muffler usually results in a loss of performance.

Chapter 8

So You Want to Start a Shop

Let's start this chapter with some general questions:
1. Are you a good mechanic?
2. Do you have operating capital, and in addition, do you have a year's salary in the bank to support your family?
3. Do you get along well with people?
4. Do you have any experience in being your own boss?
5. Are you a problem-solver?
6. Will the location provide enough business?

If you can't answer a resounding YES to at least four of these questions, perhaps you'd better think more than twice about going in any business, let alone engine repair. I don't want to discourage you—ambition and risk-taking made the American system work. But it is a fact that 9 out of 10 small businesses go under in the first 18 months because of NO's to the above six basic questions.

Question number three, "Do you get along well with people?" is perhaps the most critical. It's easy for a busy mechanic to forget that he is in business to serve people. This is not quite the same thing as repairing engines. Engine repair is part of it, but the main object of any business is to make the public happy or at least keep them satisfied.

Dealing with the public takes great globs of tack, understanding and patience. People are not rational at all times, and especially when their machines fail to function. Every customer needs, in addition to a skilled and honestly priced repair job, some modicum of emotional support. He needs your reassurance.

The worst thing that a mechanic can tell a customer is that his engine failed because he abused it. The customer is already upset, and to blame him only makes him angry. Another mistake is to promise something that you cannot deliver. You cannot make an old engine run like a new one, nor can you upset your job schedule to please everyone who is in a hurry. Nor can you be expected to work for nothing.

As a businessman you will need to keep accurate records. Discuss this matter with an accountant and with your parts supplier. Investigate the matter of insurance with a broker and keep in touch with your local banker. He can help by extending credit and by warning against some of the obvious pitfalls which beset every new business.

WORK AREA LAYOUT

Take a look at the small engine repair shops in your area. Many of them are dark, hole-in-the-wall affairs, with a crew of overworked mechanics tripping over extension cords. In these shops, the benches and the floors are spotted with grease, and old parts are stacked in the corners. The mechanics spend considerable time finding tools and as often as not, they'll depend on their "junk inventory" for replacement parts. Customers wander around at will, getting into the mechanics' way and risking injury. Prices will be set according to whatever the owner thinks the customer will pay. No records are kept of what work has been done.

True, these hole-in-the-wall operations can get the work out. Mechanics who work in these places soon learn to find the problem and to fix it. If an engine has a clogged main jet and a doubtful ignition, the jet will be cleaned. The ignition can wait for the next time. After all, the engine runs and that's all the customer can expect. Or is it?

If you are going into business, it will pay to spend some time and thought on your shop layout. In the long run an attractive, efficient work space will more than pay for itself.

The physical layout of the shop depends upon many factors—the size of the available building, the expected volume of business, the capital available, etc. But with any repair shop, whether in the home basement or in a large dealership, there are some important design factors.

Traffic Flow Considerations

When designing your shop, think of the work coming in as traffic flow. For example, if you are repairing lawnmowers, the first step is to remove grease and grime from the machine. This is a dirty job and may involve steam or high pressure water. Obviously there should be some physical separation between this stage and the precision work involved in rebuilding carburetors and ignition systems. Basic engine and transmission work demand extreme cleanliness and should be carried on in the most remote part of the shop. Some automobile dealers have a special dust-free room for this kind of work. The final stage is running adjustments and testing. Again this should be done in a separate area with very good ventilation to remove smoke and carbon monoxide. Ideally the testing should be remote from the rest of the shop. The less mechanics are distracted by noise, the more efficient they are. Some consideration should be given to soundproofing.

Timesavers

Another factor which is often overlooked is the time the mechanic spends in moving around the work area. Traditionally, shops have been arranged poorly, with benches along the walls and with a separate storeroom for the parts inventory. It is much better to locate the mechanics centrally, on individual benches in the middle of the floor. The walls can be shelved for parts storage and for rarely-used tools. Much effort can be saved if some investment is made in castered tables and dollys to move the engines from point to point about the shop. Another advantage of working on a table, rather than on the floor, is that the mechanic spends less time reaching and bending. Some large motorcycle dealers have found that an airlift, similar to the kind used to raise automobiles at a filling station grease rack, pays for itself. The mechanic works at a convenient height and the rack can be spun 360 degrees.

Parts Inventory

Parts inventory should be organized by brand and by number. The kind and amount of parts to purchase is usually suggested by the distributor. Some parts manuals have a code which indicates rapidly moving items. But it is impossible to have all the parts, all of the time. For this reason, most shops keep a collection of junk engines for pirating. The customer is then charged half the price of a new part.

Every shop should have a collection of manuals and of reprints such as provided by H.W. Sams (1014 Wyandotte St., Kansas City, Mo. 64105). Keep the old manuals because sooner or later you will need them.

Lighting

Another aspect to consider in the design of your shop is lighting. This is one factor that you can modify easily, even in a rented building. You should use as much daylight as possible, but daylight alone is rarely enough. When you need artificial light, the incandescent type is a good choice for local, high-intensity uses such as over a machine tool. Fluorescent is the best choice for lighting large areas. The tubes are installed on the ceiling, and unless the ceiling is particularly high, are shielded with covers. Unshielded fluorescent lamps give off about 4 candle power per square inch, which is a high value that causes eye discomfort if viewed directly.

The amount and distribution of light should be carefully considered. All areas of the shop do not have to be illuminated equally. The old parts bin does not require as much light as the drill press table or any other place where close work is done. These areas should be adequately, but not excessively, illuminated. It takes four times as long to perceive a small detail under 2-foot-candlepower than it does under 100-foot-candlepower. In many cases, the best solution is to mount a small auxiliary light source over the work area.

Another factor is the direction of the light. Most repair work is done on the horizontal plane and a simple overhead fixture with the light shining directly downwards may be adequate. But some work, such as adjusting carburetors or timing motorcycle ignitions, is done on the vertical. To perform these tasks efficiently, light must come in from an angle. And the amount of light is reduced whenever a person enters or a machine is brought into a room. A shop which was once well-lit will become dim as machine tools, repair jobs, and other clutter accumulate. The contrast between very dark areas and intensely light spaces can cause the individual a momentary vision loss as he moves from one area to another.

In addition, there is the matter of the quality of light. Workers object to glare, either from the source or reflected from shiny objects, extreme variations in brightness, and dense shadows. Glare can be controlled by shielding any artificial lights that are in the line of sight, by using low gloss (but not mat) finishes, and by laying out the work area so that natural light enters from above and behind. The typical shop layout with the benches facing windows is completely wrong. Extreme variations in brightness cause the eye to be distracted. It is almost impossible for the eyes not to be drawn to a bright light source. Either these bright sources must be dimmed, or the level of lighting in the whole area increased.

While it is difficult to suggest a formula for all shops, much study has been done on the proper illumination of work tables. Most modern plants use four or more rows of fluorescent lamps running parallel to the bench. The lamps are mounted high enough so that they are out of the direct line of vision of the worker, and far enough apart so that light strikes vertical surfaces.

Colors are nearly as important as the lighting. White or near-white paint should be used on the ceiling. Avoid glossy finishes as they will increase the glare and the discomfort. The walls should be of some neutral color such as light rose, buff, or slate gray. The floors should be somewhat darker. The idea of painting floors to be reflective is probably a mistake. All that is needed is enough reflection to find small parts when they are dropped. Machines should always be painted in colors that contrast with their surroundings.

SAFETY IN THE SHOP

Many mechanics are needlessly hurt every year because they did not follow elementary safety precautions. Here are minimum safety considerations. More information can be obtained from an agent who specializes in industrial insurance coverage.

Grinders

Bench grinders are indispensable tools for every shop, but can be very dangerous if not used properly. Each abrasive wheel has its maximum allowable rpm marked on the hub. In general, the larger the diameter of the wheel, the lower the rpm. Most bench grinders run at 3,600 or 4,000 rpm. Never use a lower speed wheel, since there is a very real possibility that the wheel will explode. All grinders, even the portable models, should be fitted with a heavy steel guard around the perimeter of the wheel. These guards sometimes get in the way when grinding large objects, but they should not be removed. The typical Type I wheel with straight sides is strong only on its outer circumference. Grinding on the sides of the wheel will overstress it and cause breakage. If the wheel begins to vibrate step out of the way and cut power to the machine. The wheel has cracked or else the shaft has bent.

Electrical Tools

All portable electrical tools should have a three-wire grounding system. If a drill motor or other tool seems to be overheating, take the load off the tool and run it a minute or so. The fan or self-cooling of the tool will cool the armature and the fields. Do not force a hot motor since the insulation will melt and the tool will short out. Half-inch size electric drills develop enough torque to break an arm. Be sure that you have a firm, double-handed grip on the tool so that you can control it if it stalls.

Welders, Torches, and Hazards

Electric arc welders should have a separate, fused circuit. These machines draw 230 volts on the primary side, which is stepped down to between 25 to 30 volts across the electrodes. Because of the low voltage output, they are not considered particularly hazardous. However, men have been killed welding in wet weather and on those rare occasions when the machine malfunctions. All of these welders have a duty cycle, marked on the case, that must be adhered to. The duty cycle generally runs between 10 and 20 percent. This means that the welder must be allowed to cool between 80 and 90 percent of the time. Roughly, use it for 6-12 minutes out of each hour, then let it cool.

All electric welding should be done in a well-ventilated, lightproof stall. The radiation given off by the arc will destroy the retina of the eye; so wear a welder's helmet or goggles. Also, the radiation will produce severe burns on unprotected skin, similar to sun burn, so wear heavy clothing.

Oxy-acetylene torches are preferred over electric welders for small shops because of their versatility. These torches can be used to bend, form, shrink, weld, braze, and cut. But there are real hazards associated with them. Acetylene is a highly explosive gas which should be treated with the greatest respect. At a 2.5 percent proportion with air, this gas is explosive, and it continues to be explosive until the proportion reaches 82 percent. No other industrial gas has such a wide explosive range. At pressures above 15 psi, acetylene will explode by decomposition without the presence of air. Oxygen will spontaneously ignite in the presence of oil and grease. The hoses, torch handles, and the regulators must be kept free of petroleum products.

Before using the equipment, inspect it for cleanliness and for leaks. Hoses cannot be safely repaired: when they show signs of deterioration, they should be replaced. The bottles should be stored upright out of the sun, and the regulators periodically returned to the distributor for inspection. Unless you are experienced in this line of work, do not attempt to repair or to make internal adjustments on the regulators yourself. If you suspect a leak in the system, make a bubble test with Ivory soap. DO NOT USE ANY OTHER BRAND OF SOAP BECAUSE OF THE DANGER OF OXYGEN COMBINING WITH IT.

When preparing to use the torch, make certain that the regulator valves are screwed all the way out to the "off" position before the main tank valves are opened. This is to protect the regulators from the sudden impact of tank pressure. And when opening the tank valves, stand alongside of the regulators, out of the way, in case they blow out. Backfiring or "machine gunning" at the torch is very dangerous and can lead to a major explosion.

Wherever possible, welding should be done in one location, well away from inflammables. Some shopes construct a steel table, topped with firebrick. The table can also be fitted with a vise and should have a water trough for quenching.

Cleanliness

Rubbish, scrap paper, and litter are highly combustible. Such material should be stored in metal containers well away from sources of low heat, such as steam pipes, and should be entirely clear of sparks and flame. If there are oily wastes, such as greasy lunch papers or packing from parts cartons, spontaneous combustion is possible. Remove these wastes daily to some point outside of the building, and preferably store them, until pickup, in covered containers.

Oily rags are part of repair work. They are especially dangerous if they have been wetted with gasoline. Do not attempt to wash and re-use these rags. The risk is not worth it. Many shops subscribe to an industrial towel service which picks up the rags once a week or so. The rags should be stored in a closed, fire-resistant container, and preferably outside of the building.

Wiring

All wiring should conform to the National Electrical Code and whatever local codes apply. See that the fuses are not over-rated, and check the fuse connections for tightness and good contact. Dimming of the lights when power equipment is on the line is a sure sign of overloaded wiring. Electrical wires do not have to turn red hot to cause fires. A relatively low temperature over an extended period is just as dangerous. Motors with open cages should be blown out regularly to prevent dust accumulation. Universal motors of the kind used on portable power tools should be repaired when the brushes show signs of excessive arcing.

Handling Gasoline and Solvents

The big danger in a repair shop is gasoline. It will explode when the proportion of vapor to air is between 1.4 and 6 percent. At lesser or greater proportions, there will be no combustion. But at 2.25 percent, the explosion will be particularly violent and will have the burst energy of an equivalent weight of TNT. If mixed with air in the proper amounts, the fuel in the tank of a small outboard can literally blow a building apart.

Gasoline vapors are heavier than air and will travel along the floor for long distances. When ignited the flame will trail back to the source and ignite it. One gasoline fire on record was ignited 162 feet from the container.

There are precautions in the handling and storage of gasoline which greatly reduce the chance of fire. Always use an approved container having a spout and a snap cover. The container should be located well away from power cutting tools and from welding equipment. When refueling an engine, ground the spout to the side of the tank to prevent the build up of static electricity. The practice of straining gasoline through a chamois is dangerous because it encourages sparking. Never refuel an engine that is running or one that is hot. A hot exhaust pipe can ignite the fuel.

Engines which are in the shop for repair should be thought of as fuel tanks and should be stored in a well-ventilated place. Any tank which shows evidence of leakage should be drained immediately.

Never use gasoline for cleaning purposes. Use kerosene; it dissolves grease quicker and does not leave a residue of varnish and gum. As a general rule, the use of air guns to apply solvent should be avoided. These devices are convenient and relatively fast, but the fog of solvent is dangerous and unhealthful. A better method is to use steam for large surfaces and a mixture of Gunk and kerosene for the smaller parts.

And finally, keep the total amount of flammables in the shop at the absolute practicable minimum.

Fire Extinguishers

All shops should be equipped with fire extinguishers located at critical places such as the welding room and the fuel storage area. Underwriters' Laboratories classify fires as follows:

Class A—ordinary materials such as paper, wood, and packing, where the cooling effect of water and of water-based solutions is of primary importance.

Class B—flammable liquids, greases, etc., which must be extinguished by a blanket effect to cut off the oxygen.

Class C—electrical equipment, where the use of a non-conducting extinguishing agent is mandatory.

Unfortunately, a universal extinguisher, equally good for all classes of fire, has yet to be developed. The soda-acid type is excellent for Class A fires, but must never be used on Class C electrical fires, because the solution in these extinguishers is highly conductive. Oil and grease fires can be extinguished with it, but the solution has a water base and tends to spread the fire by floating the burning liquids.

Carbon dioxide extinguishers are the type most frequently found in small shops. They contain CO_2 pressurized to 800-900 pounds per square inch. When aimed at the base of a fire, the carbon dioxide will form a dense blanket around it. The fire is primarily extinguished from lack of oxygen, aided by the cooling effect of the expanding gas. These extinguishers are recommended for Class B and C fires and can be used to put out Class A fires. While most CO_2 bottles have a pressure gauge on the handle to show the state of charge, the only sure method is to weigh the bottle. A leak-down rate of 10 percent per year is accepted as normal.

Dry chemical extinguishers are also quite effective for Class B and C fires. The agent is powdered bicarbonate of soda. It is chemically processed to make it waterproof and to improve its flow characteristics. When using these extinguishers, it may be necessary to tap the horn to get the bicarbonate to flow.

Carbon tetrachloride extinguishers are no longer recommended and should be discarded. They work well on Class B and C fires but are hazardous to the operator. Carbon tetachloride fumes are dangerous to inhale at room temperatures. When heated, carbon tet gives off highly toxic gases such as hydrochloric acid and phosgene. The potency of phosgene can be gauged by the fact that it was used as a poison gas during the First World War. If you are compelled to use one of these extinguishers, empty it on the fire and leave the room immediately.

CUSTOMER RELATIONS, AND TRANSACTION RECORDS

Customers should be kept out of the work area. They distract the mechanics and can be a general nuisance. Some of them offer technical advice, and are offended when their advice is not followed. Another reason for keeping them out is the matter of liability. If a customer is hurt, the owner is liable for real damages and whatever punitive damages the court may decide. Today, personal injury judgments can reach six figures.

Most shops separate the customer waiting room and the work area proper with a counter. The waiting room should be furnished with chairs, current magazines, and coffee. New engines, parts, and accessories can be displayed in glass cases. It is a good idea to have a place showing the list of charges for routine jobs. This will reassure the customer that your shop is not a "gyp joint" and will give him some idea of costs.

Estimating

Customers expect a cost estimate when they bring their machine in for service. This is only fair. The customer should know exactly what work you plan to do and how long it will take. The estimate should have a 10 percent or so over-charge built into it, so that there is some room to replace items that you may have overlooked. Wherever possible, sell the customer on a standard job. Suppose the ignition is out and you give him an estimate of $9 labor and $4.50 parts, to replace the points and condenser. This is fine—if only the ignition is out. It is possible that the carburetor or the timing, or the generator brushes or a dozen other components have also failed. So it is much better to estimate a complete tune-up. If he just wants the engine to run, explain that there might be other things wrong with it, but do what he says.

Keeping Track of the Job

A job sheet should be made out on each repair item. The sheet should be dated, contain the customer's full name, his home address, phone number, and the serial number of the machine. The reason for service should be clearly spelled out, along with the estimate and any particular instructions the customer might have. The mechanic should list all the parts he replaced, and the total in labor and parts charged. This sheet is then filed for reference. The customer can be put on a mailing list for promotional material. And if he is dissatisfied with the service, there is some record of what was done.

Complaints

"Call backs" should be repaired immediately and preferably by the mechanic who did the work the first time. Try not to charge the customer on call backs. True, most of the time the fault will lie with him. He will have used contaminated fuel, or upset the carburetor adjustments, or will have carelessly damaged the engine in some other way. It is easy to become morally indignant, especially since the customer is probably already angry. But it is better to let him have the free service. Otherwise, you're almost certain to lose his business.

If you hold that the customer is always right, you may lose money on him today, but over the years you will make money.

SHOP TOOLS AND SUPPLIES

If you're working on your own equipment or just starting out in business, it would be a mistake to invest heavily in tools. Most mechanics start out with a basic inventory and add special tools as they need them.

Basic Tools

Here's a list of tools that are the basic essentials for working on small gas engines.

1. One 8 oz. machinist's hammer.
2. Assorted screw drivers, flat and Phillip's No. 1 and No. 2.
3. A set of open-end wrenches $3/8$" to 1". And if you work on Japanese and European equipment, add 4 mm to 19 mm open-end wrenches.
4. A socket set, $3/8$" square drive in the size range listed in Item 3. Some mechanics use the smaller $1/4$" drive and feel better at the end of the day for it.

5. A ratchet handle.
6. A spark plug socket (either a special socket for this purpose or a 13-16" deep well).
7. One 8" adjustable wrench. For durability, buy the best quality that you can find such as Blackhawk or Snap-on.
8. A pair of curved-jaw Vise-Grips.
9. A set of E-Z outs for removing broken fasteners.
10. Allen wrenches, American or Metric (the latter is available from K-D Tools).
11. An 8" mill bastard file.
12. Emery cloth, 1-inch strip, medium grit.
13. Two parts cleaning brushes, one a wire bristle type, and the other a camel's hair brush.
14. An electric drill motor, preferably $3/8$" chuck, although you can get by with a $1/4$".
15. A set of high-speed drill bits.
16. Feeler gauges.

Additional Tools

As you get deeper into repair work, the tools listed below will become necessary.

1. A pair of long-nosed pliers.
2. A pair of diagonal side-cutters (often known as "Dykes") for cutting wire and removing stubborn shaft keys.
3. An inexpensive multi-meter which should have an accuracy of at least 10,000 ohms to the volt on DC. If you buy a Japanese unit, the first thing to do is to make up a set of longer probe leads for it. One probe should have a miniature alligator clip soldered to its tip.
4. A soldering gun of at least 250 watt rating.
5. A $3/8$" square-drive impact driver with Phillips and Posidriv No. 2 and No. 3 recess bits. When struck with a hammer, the driver will turn a few degrees, loosening the most stubborn screw. This tool is vital when working on power chain saws and on Japanese motorcycles.
6. An 8" gear puller.
7. A torque wrench. There are many kinds of torque wrenches available, but from the point of view of reliability and economy, the moving-arm type (such as sold by Sears) is the best. A 0 to 300 pound-inch type is adequate for all jobs, although a smaller wrench is sometimes more convenient.
8. A piston ring compressor which goes down to 2" bores. The Briggs and Stratton part number 19070 works well.

9. A valve spring compressor for small engines (Briggs and Stratton No. 19063).
10. A suction cup tool (K & D No. 501).
11. A compression gauge, preferably the screw-in type for one-hand operation.
12. A blade balancer for lawnmower work (available from Sears).
13. Tap-Lok (for Class 2 metals) or Heli-Coil Master Thread Repair Pack.

Nonspecialized Shop Tools

In addition, you will need some heavy shop tools. They are listed here in a rough order of utility.

1. A brass-jawed vise, the heavier the better.
2. A bench grinder with a wire wheel on one side. It should develop at least 0.3 horse power, and have a half-inch shaft and sealed ball bearings.
3. A source of compressed air which must be regulated not to exceed 30 pounds per square inch. WARNING: AT MORE THAN 30 PSI AIR CAN PENETRATE HUMAN TISSUE AND ENTER THE BLOOD STREAM.
4. A welder. Many shops have "cracker box" electric arc machines, but if you are going to spend the money, an oxy-acetylene torch is much more versatile.
5. An arbor press for removing bearings and which may be used, with a dial indicator and vee-blocks, for straightening crankshafts (although manufacturers frown on this). You can construct a press using a 5-8 ton hydraulic jack and half-inch steel plate.

Supplies and Materials

1. Kerosene or mineral spirits for cleaning. NEVER USE GASOLINE FOR CLEANING!
2. Carburetor cleaner (available as Bendiz in 5-gallon tins or as Gunk in pint containers).
3. Gasket cement such as Permatex Aviation No. 3.
4. Silicone cement, G.E. Silicone Seal or equivalent.
5. Epoxy cement.
6. Locquic Grade T Primer, and Loc-Tite Sealant.

7. Bearing assembly grease.
8. Solder, acid and rosin core.
9. Masking tape.
10. Two-cycle oil (such as refined by Texaco or Shell).
11. Four-cycle oil (at least MS grade).
12. A supply of fresh fuel (most small engines operate on unleaded gasoline with a minimum octane rating of 90 but a few high-performance motorcycles require a higher octane). Do not use fuel containing carburetor cleaner in two-cycle engines. The cleaner will remove the oil.

A FINAL WORD

Let's assume that you're successful. Everything has jelled, money's pouring in; you have a good crew, many customers, a good reputation, etc. You also have created a valuable target for thieves and vandals. So just as soon as you can, have your shop wired in to an automatic fire-burglar alarm system, preferably one that dials the police.

Index

A

Accessories and controls	235
Adjustable backlash	225
Adjusting the diaphragm-type carburetor	111
Adjusting the float-type carburetor	103
Adjusting the suction-lift carburetor	95
Advance mechanism	69
Air filters	119
Air vane type governor	257
Alternating current systems	140
Alternators	141
Analysis and inspection	160
Arms, rocker and push rods	171
Automatic chokes	114

B

Backlash, adjustable	225
Ball and cam clutch	235
Basics, gas engine	7
Batteries, lead-acid	149
Batteries servicing	151
Battery chargers	151
Battery ignition system	69
Battery ignition system, further tests	71
Battery ignition system troubleshooting	71
Bearings, big end	191
Bearings, main	197
Bearings, small end	194
Belt tension	209
Big end bearings	191
Bore and stroke	7
Breaker points	63
Briggs composite carburetors	112

C

Cable and rope	245
Cable construction, high tension	60
Cables, high tension	60
Cables, testing	61
Cam and ball clutch	235
Camshafts	200
Capacitive discharge systems	81
Carburetion, principles	91
Carburetors	91
Carburetors, diaphragm-type	107
Carburetors, float-type	95
Carburetors, multiple	113
Carburetors, suction-lift	93
Centrifugal type governors	258
Chain drive	211
Chain stretch	212
Chain wear	213
Chargers, battery	151
Checking valve wear	165
Checks, compression	40
Checks, fuel system	39
Checks, ignition	35
Chokes, automatic	114
Chokes, repair and replacement	115
Chokes, testing	117
Circuit breakers and fuses	149
Classification of engine by speed	15
Cleaning the diaphragm-type carburetor	111
Cleaning the float-type carburetor	101
Cleaning the suction-lift carburetor	94
Clutch, ball and cam	235

Clutches	229
Coil construction	49
Coil polarity	51
Coils	49
Coil testing	51
Combination, starter-generator	135
Composite carburetors, Briggs	112
Compression checks	40
Compression ratio	7
Compression releases	261
Condenser construction	64
Condenser faults	65
Condensers	64
Condenser, troubleshooting	65
Connecting rods	189
Construction coil	49
Construction condenser	64
Construction, spark plug	55
Controls and accessories	235
Controls, remote	253
Coolers, oil	267
Cooling	19
Crankcase	194
Crankshaft	198
Current and voltage regulators	143
Current generators, direct	127
Current systems, alternating	141
Cylinder head, removing	159
Cylinder head, replacing	163
Cylinder heads	158
Cylinders	185
Cylinders, number and placement	11

D

Detailed inspection and testing	77
Determining the problems	176
Diaphragm-type carburetor cleaning	111
Diaphragm-type carburetors	107
Direct current generators	127
Discharge systems, capacitive	81
Distributors	67
Drive, chain	211
Drive, friction	207
Drive ratios	212
Drives, gear	213
Drive, shaft	205
Drive, v-belt	208
Dunking (immersion in water)	44

E

Electrical systems	123
Electrical theory	123
Electric shift, OMC	226
Electronic regulation	147
Energy transfer systems	79
Engine classification by speed	15
Engine, gas, basics	7
Engine service	157
Equipment, test	125
Erratic idle	43
External magnetos	79

F

Failure to start when hot	42
Faults, condenser	65
Filters, air	119
Finding the timing marks	69
Float-type carburetor, cleaning	101
Float-type carburetors	95
Flywheel, pulling	75
Four-cylinder operation	28
Friction drive	207
Fuel	26
Fuel lines	87
Fuel pumps	87
Fuel stoppage	90
Fuel system checks	39
Fuel systems	85
Fuel tanks, leaks, and repair	120
Further tests, battery ignition system	71
Fuses and circuit breakers	149

G

Gas engine basics	7
Gear cases, outboard	221
Gear drives	213

General troubleshooting procedure	86
Generators, direct current	127
Generator-starter combination	135
Governors	257
Grinding, cylinder	182

H

Heads, cylinder	158
Heli-Coil, inserting	204
High tension cable construction	60
High tension cables	60
Horsepower and torque	8

I

Idle, erratic	43
Ignition checks	35
Ignition system, battery	69
Ignition systems	47
Ignition systems, magneto	73
Ignition systems, special	78
(Immersion in water) dunking	44
Impulse starters	247
Inherent problems of magnetos	73
Inserting a Heli-Coil	204
Inspection and testing, detailed	77
Inspection before overhaul	215
Inspection, lawnmower shaft	205

K

Kickstarters	229

L

Lack of power	42
Lapping valves	166
Lawnmower shaft inspection	205
Lead-acid batteries	149
Leaks, and repair, fuel tanks	120
Lifter noise	173
Lifters, valve	171
Liners and sleeves	185
Lines, fuel	87
Lubrication	22

M

Magneto ignition systems	73
Magnetos, external	79
Magnetos, inherent problems	73
Magnetos, servicing	75
Main bearings	197
Maintenance, periodic	213
Maintenance, spark plug	57
Marks, timing	199
Mechanical regulators	143
Mechanism, advance	69
Motors, starter	133
Mufflers	268
Multiple carburetors	113

N

Noise	18
Noise, lifter	173
Number and placement of cylinders	11

O

Oil coolers	267
OMC electric shift	226
Operating principles	27
Operation, four-cylinder	28
Operation, two-cylinder	31
Outboard gear cases	221
Overall troubleshooting principles	35
Overhaul	217, 224, 228
Overhaul —inspection before	215
Overheating	43

P

Periodic maintenance	213
Pistons	174
Pistons, rebuilding	177
Pistons, removing	175
Placement and number of cylinders	11
Plugs, spark	53
Point gap setting and repair	63

Points, troubleshooting	63
Polarity, coil	51
Power, lack of	42
Power transmission	205
Precautions, safety	250
Pressure testing	197
Principles of carburetion	91
Principle of troubleshooting and theory of operation	7
Principles, operating	27
Principles, overall troubleshooting	35
Probable faults	223
Problems, determining	176
Procedure, general troubleshooting	86
Pulling the flywheel	75
Pumps	263
Pumps, fuel	87
Push rods and rocker arms	171

R

Ratchets	237
Radiators	266
Ratio, compression	7
Ratios, drive	212
Reassembly	225
Rebuilding pistons	177
Rectifiers	141
Regulation, electronic	147
Regulation, voltage and current	143
Regulators, mechanical	143
Releases, compression	261
Remote controls	253
Removing pistons	174
Removing the cylinder head	159
Removing valves	165
Repair and replacement, chokes	115
Repair, and setting point gap	63
Repair, fuel tank leaks	120
Replacement, sleeve	186
Replacing the cylinder head	163
Replacing valve guides	165
Rewind starters	235
Rewiring	155
Rings	178
Rocker arms and push rods	171

Rods, connecting	189
Rods, push and rocker arms	171
Rope and cable	245
Rope-wind springs	242

S

Safety precautions	250
Sealing	178
Seals	197
Seats, valve	167
Selecting wire	153
Service	130, 227
Service engine	157
Servicing	135
Servicing batteries	151
Servicing magnetos	74, 75
Setting point gap and repair	63
Shaft drive	205
Shaft, straightening	207
Sleeve replacement	186
Sleeves and liners	185
Small end bearings	194
Solenoids	139
Spark plug construction	55
Spark plug maintenance	57
Spark plugs	53
Spark plug tests	55
Special ignition systems	78
Specific symptoms, troubleshooting	42
Speed, classification of engine	15
Spots, trouble	252
Springs, rope wind	242
Springs, valve	167
Starter-generator combinations	135
Starter motors	133
Starters, impulse	249
Starters, rewind	235
Starters, vertical pull	246
Start, failure when hot	42
Stoppage, fuel	90
Straightening a shaft	207
Stretch, chain	212
Stroke and bore	7
Suction-lift carburetor, adjusting	95
Suction-lift carburetor, cleaning	94
Suction-lift carburetors	93
Summary	46

Symptoms, specific troubleshooting	42
System checks, fuel	39
Systems, electrical	123
Systems, energy transfer	79
Systems, fuel	85
Systems, ignition	47
System troubles	87

T

Tension, belt	209
Test and inspection, distributor	67
Test equipment	125
Testing and inspection, detailed	77
Testing chokes	117
Testing, coil	51
Testing of cables	61
Testing, pressure	197
Tests, spark plug	55
Theory, electrical	123
Theory of operation, and troubleshooting principles	7
Thermostats	263
Timing	67
Timing marks	199
Timing marks, finding	69
Torque and horsepower	8
Transaxles	228
Transmission, power	205
Transmissions	214
Troubleshooting	124
Troubleshooting battery ignition system	71
Troubleshooting principles and theory of operation	7
Troubleshooting principles, overall	35
Troubleshooting procedure, general	86
Troubleshooting specific symptoms	42
Troubleshooting the condenser	65
Troubleshooting the points	63
Trouble spots	252
Two-cylinder operation	31
Type, air vane governor	257

V

V-belt drive	208
Valve guides, replacing	165
Valve lifters	171
Valves	163
Valve seats	167
Valves, lapping	166
Valve springs	167
Valves, removing	165
Valve wear, checking	165
Vertical-pull starters	246
Vibration	17
Voltage and current regulation	143

W

Wear, chain	213
When hot, failure to start	42
Wire, selecting	153
Wiring	153

623.8　　Dempsey
　　　　How to repair small gasoline
　　engines　OCT 18 1976
　　　　　　　　　　AUG 3　1981

DATE DUE

AUG 2 1995			6 1991
			1992

Demco, Inc. 38-293